Professional Users Handbook for Rock Bolting

Other volumes published within the Series on Rock and Soil Mechanics

Vutukuri, V.S. & Lama, R.D.:
Handbook on Mechanical Properties of Rocks Vols. I—IV
1974/1978

Hardy, H.R., Jr. & Leighton, F.W.:
Acoustic Emission/Microseismic Activity in Geologic Structures and Materials,
First Conference
1977

Karafiath, L.L. & Nowatzki, E.A.:
Soil Mechanics for Off-Road Vehicle Engineering
1977

Baguelin, F., Jézéquel, J.F. & Shields, D.H.:
The Pressuremeter and Foundation Engineering
1978

Assonyi, Cs. & Richter, R.:
The Continuum Theory of Rock Mechanics
1979

Hardy, H.R., Jr. & Leighton, F.W.:
Acoustic Emission/Microseismic Activity in Geologic Structures and Materials
Second Conference
1979

Hanna, T.H.:
Foundations in Tension Ground Anchors
1982

Jumikis, A.R.:
Rock Mechanics
Second Editon
1983

Hardy, H.R., Jr. & Leighton, F.W.:
Acoustic Emission/Microseismic Activity in Geologic Structures and Materials
Third Conference
1984

Hardy, H.R., Jr. & Langer, M.:
The Mechanical Behavior of Salt
First Conference
1984

Hanna, T.H.:
Field Instrumentation in Geotechnical Engineering
1985

Mazurkiewicz, B.K.:
Offshore Platforms and Pipelines
Selected Contributions
1986

Farouki, O.T.:
Thermal Properties of Soils
1986

Hardy, H.R., Jr. & Langer, M.:
The Mechanical Behavior of Salt
Second Conference
1986

Stacey, T.R. & Page, C.H.:
Practical Handbook for Underground Rock Mechanics
1986

Series on Rock and Soil Mechanics
Vol. 15 (1986)

PROFESSIONAL USERS HANDBOOK FOR ROCK BOLTING

Bengt Stillborg
Tek. lic., Dr. Eng.

1986

TRANS TECH PUBLICATIONS

Professional Users Handbook for
Rock Bolting

Series on Rock and Soil Mechanics
Vol. 15 (1986)

Exclusive Worldwide Sales &
Distribution Rights
by
TRANS TECH PUBLICATIONS
P.O. Box 1254
D-3392 Clausthal-Zellerfeld
Federal Republic of Germany

ISBN 0-87849-073-6

Layout and phototypeset
by schriftsatz-studio kurt grohs · D-3401 Landolfshausen

Printed and bound in the Federal Republic of Germany

Foreword

Rockbolting is the most effective and the most economical means of supporting excavations in rock in both mining and civil engineering applications. Rockbolts can be used to control all types of instabilities except those involving extremely weak and soft ground such as that which may occur in a major fault zone. Various types of bolts, both tensioned and untensioned, are used today and an understanding of the way in which these bolts work is essential if they are to be used correctly.

Atlas Copco, as a major mining equipment manufacturer, has been involved in the practical aspects of rockbolt applications for many years. With the introduction of their Swellex bolting system, they are also now suppliers of rockbolts. This puts them into an ideal position to prepare a users handbook in rockbolting and I am delighted that they have recognised and accepted this challenge. The text which follows is a sound practical handbook which reviews the basic principles of rock bolting and sets out the design considerations used for most types of rockbolts in current use. It discusses the characteristics of these bolts and goes on to give a wealth of information on installation procedures and the observations and measurement of rockbolt performance. It is written with the professional miner or civil engineer or contractor in mind and these users should have a clear and uncluttered presentation of the information which they require in order to make rational design and job site decisions.

I commend Dr. Bengt Stillborg and Atlas Copco for their efforts in preparing this handbook and I look forward to using it in my own consulting and educational activities in rock engineering.

Vancouver
November, 1985

Dr. EVERT HOEK

Acknowledgements

This handbook was initiated as a result of a visit to a tunnel work site beneath the harbour of Copenhagen. Mr. Lorne Herron, from Atlas Copco Canada, who has extensive experience of practical aspects of rock reinforcement, expressed the need for a users handbook for rockbolting. The task of writing the handbook was later given to the undersigned, with the objective that the book should cover aspects of rockbolting from site investigation through design, selection, installation, tests, monitoring and costs of rock reinforcement.

The financial support provided by Atlas Copco MCT AB is greatly appreciated.

Special acknowledgements are directed to Mr. Robert Robertson, Mr. Björn Kåreby and Mr. Lorne Herron, all at Atlas Copco MCT, for their vital role in organizing the support received from Atlas Copco. Their assistance and encouragement is gratefully acknowledged. Special acknowledgements are also directed to Professor Ove Stephansson for his keen reading and suggested improvements to the manuscript, and for valuable discussions. The author and sponsors of this handbook extend a special thanks to Dr. Evert Hoek for his review and comment on the work. I would also like to thank Mr. Arne Nyström for his preparation of the illustrations.

Luleå
March, 1986

Bengt Stillborg

Contents

PART II*

* authored by representatives of Atlas Copco MCT.

Terminology

Rock reinforcement	The installation of rockbolts, cables or any other type of element in a rock mass to reinforce and mobilize the inherent strength of the rock, so that the rock becomes "self supporting". The rock reinforcement element is installed inside the rock mass, that is, it forms part of the rock mass.
Rock support	The application of a concrete lining, steel sets or any other type of "engineering" structure which will restrict movements of the rock. Rock support is external to the rock mass.
Rock stabilization	Combined application of rock reinforcement and rock support to prevent failure of the rock mass.
Pre-reinforcement	Installation of reinforcement in a rock mass before excavation commences.
Rockbolts	A reinforcing element, normally made of solid or tube formed steel installed untensioned or tensioned in the rock mass.
Cable bolts	A reinforcing element, normally made of steel wires which are layed to a strand or a rope configuration and installed untensioned or tensioned with cement grouting in the rock mass.
Rock mass	In-situ rock, composed of small or large pieces of solid rock limited by discontinuities.
Failure of a rock mass	Loosening or fall-out of rock from the rock mass.

Conversion Factors

Physical quantity	Symbol			
Length	(m)	1 in = 0.0254 m 1 ft = 0.3048 m 1 yd = 0.9144 m 1 mile = 1609.344 m	1 m =	39.3701 in 3.28084 ft 1.09361 yd $0.621371 \cdot 10^{-3}$ mile
Area	(m^2)	1 in^2 = 645.16 mm^2 1 ft^2 = 0.092903 m^2 1 yd^2 = 0.836127 m^2 1 acre = 4046.86 m^2	1 m^2 =	1550 in^2 10.7639 ft^2 1.19599 yd^2 $0.247105 \cdot 10^{-3}$ acre
Volume	(m^3)	1 in^3 = 16.3871 ml 1 ft^3 = 28.3168 l 1 yd^3 = 0.764555 m^3 1 UK gal = 4.54609 l 1 US gal = 3.78541 l	1 l =	61.0237 in^3 $35.3147 \cdot 10^{-3}$ ft^3 $1.30795 \cdot 10^{-3}$ yd^3 0.219969 UK gal 0.264172 US gal
Mass	(kg)	1 lb = 0.453592 kg 1 oz = 28.3495 g 1 ton UK = 1016.5 kg 1 ton US = 907.185 kg	1 kg =	2.20462 lb 35.274 oz $0.984207 \cdot 10^{-3}$ ton UK $1.10231 \cdot 10^{-3}$ ton US
Density	(kg/m^3)	1 lb/ft^3 = 16.0185 kg/m^3 1 lb/in^3 = 27679.9 kg/m^3	1 kg/m^3 =	0.0624278 lb/ft^3 $36.127 \cdot 10^{-6}$ lb/in^3
Force	(N)	1 kp = 9.80665 N 1 lbf = 4.44822 N	1 N =	0.101972 kp 0.224809 lbf
Moment	(Nm)	1 kpm = 9.80665 Nm 1 lbf · ft = 1.35582 Nm	1 Nm =	0.101972 kpm 0.737562 lbf · ft
Pressure or stress	(Pa)	1 bar = 100 kPa 1 kp/cm^2(at) = 98.0665 kPa 1 psi = 6.89476 kPa	1 kPa =	0.01 bar 0.0101972 kp/cm^2(at) 0.145038 psi
Energy	(J)	1 kWh = 3.6 MJ 1 kpm = 9.80665 J 1 kcal = 4.1868 kJ 1 hkh = 2.6478 MJ	1 kJ =	$0.277778 \cdot 10^{-3}$ kWh 101.972 kpm 0.238846 kcal $0.377673 \cdot 10^{-3}$ hkh

SI-unit prefixes

Prefix	tera	giga	mega	kilo	milli	micro	nano	pico
Symbol	T	G	M	k	m	μ	n	p
Multiplier	10^{12}	10^{9}	10^{6}	10^{3}	10^{-3}	10^{-6}	10^{-9}	10^{-12}

Chapter 1

Introduction

The history of rockbolting dates from the end of the 19th century. However, rockbolts were not used extensively until forty or fifty years later. Since then, the use of rockbolts in mining as well as in underground excavations for civil engineering applications has become worldwide and hundreds of millions of bolts are installed annually.

In recent years the range of applications for rockbolts has widened due to advances in rock mechanics and the increasing use of rock reinforcement in underground excavations, as an alternative to more traditional forms of support. Also, the development of new rockbolt concepts has lead to the use of rockbolts in non-traditional applications.

The engineer who must design rock reinforcement systems today is faced with an increasing demand to optimize his design with respect to both safety and economic considerations. This handbook attempts to summarize state-of-the-art of rockbolting procedures, to guide engineers and operators in rock reinforcement problems. It is also hoped that the book will be useful to both research and teaching institutions, as well as companies undertaking work in rock reinforcement.

The increasing use of rock reinforcement in underground excavations has resulted in a number of publications on rock reinforcement. This book contains a review of many of these publications, response to questionnaires and personal inquiries, as well as the professional experience of the author. It covers most aspects of rock bolting from a presentation of the most commonly used types of rockbolts through installation, site investigation, choice of rockbolt system, design, test and monitoring of rock reinforcement and cost. Also included are some aspects of cable bolting.

The variety of rockbolts used at different work sites worldwide is almost unlimited, as is the variety of the rock conditions in which bolts are used. One soon realizes in writing a users handbook for rock bolting, the impossibility of satisfying the requirements of all possible users. For clarity, and to keep the contents of this book within reasonable limits, it has been necessary to restrict the number of rockbolts described. The variety of rock conditions considered for design of rock reinforcement is also restricted to those of general interest.

Chapter 2

Why Rockbolts?

The principal objective in the design of a support system is to help the rock mass to support itself. This applies to rock reinforcement systems, e.g., where the rockbolts actually form part of the rock mass, as well as to rock support systems e.g. steel sets which do not form part of the rock mass but support it externally. The rockbolt reinforces and mobilizes the inherent strength of the rock mass. Steel set or shotcrete support systems, however, restrict movements of the rock mass externally. Both systems can be used for both temporary as well as permanent rock mass support.

Rockbolts are the most commonly used system for rock mass stabilization, particularly in the mining industry.

There are a number of reasons for the widespread use of rockbolt reinforcement systems. Some of these are as follows;

— versatile, can be used in any excavation geometry
— usually simple to apply
— relatively inexpensive
— installation can be fully mechanized.

In addition, after blasting and mucking, prompt installation of rockbolts is very important from a rock mechanics point of view. Using rockbolts, the reinforcement density (spacing between bolts and bolt length) can be modified. This is frequently required by local rock mass conditions. Another advantage is that rockbolts can easily be combined with additional support systems such as wire mesh, shotcrete or a concrete lining. In many methods for underground excavation and support like the New Austrian Tunnelling Method (NATM) systematic rock bolting is combined with shotcrete and for permanent reinforcement of the rock, a concrete lining.

A flow chart for a rockbolting program is illustrated in Fig. 1. It is important that the overall control of the rockbolting program (all the individual activities included in the program) is in the hands of a single engineer.

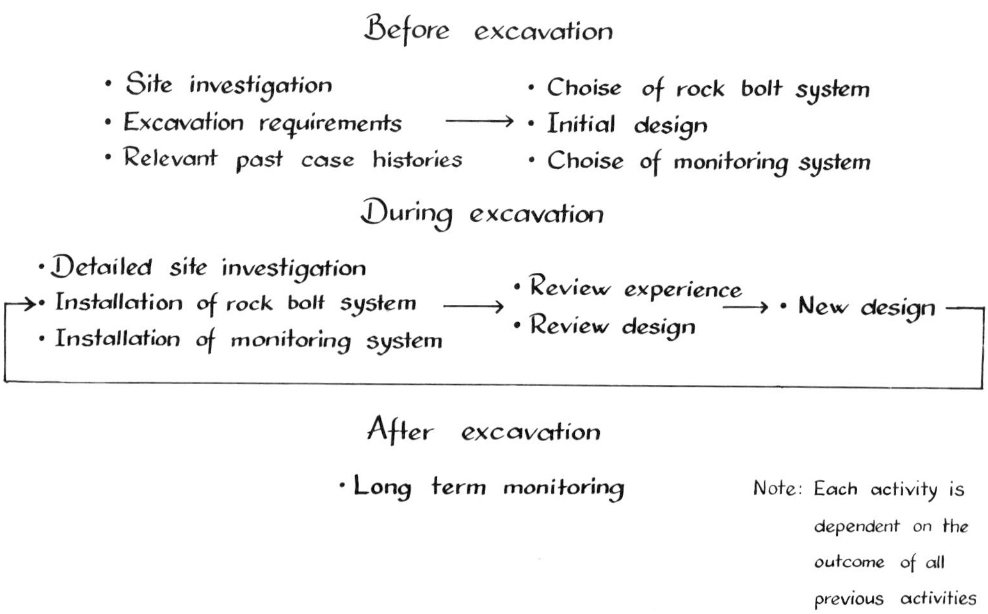

Fig. 1 Flow chart for rockbolting.

Chapter 3

Review of Typical Rockbolt Systems

A number of different types of rockbolts are now used worldwide. Many rockbolt types show only minor differences in their design and are basically varieties of the same concept. In a review of typical rockbolt systems, it is therefore possible to arrange the different types of rockbolts in groups and present a representative for each group. The groups are made with reference to the different anchoring techniques. Only the most widely used rockbolt type from each group will be considered.

The following groups of bolts are considered:

— mechanically anchored rockbolts
— grouted rockbolts
— grouted cable bolts
— friction anchored rockbolts.

3.1 Mechanically Anchored Rockbolts

The expansion shell anchored rockbolt, of standard or bail type, is the most common form of mechanically anchored rockbolt. The expansion shell anchor operates basically in the same manner whether it is of standard or bail type. A wedge attached to the bolt shank is pulled into a conical expansion shell as the bolt is rotated. This forces the shell to expand against and into the wall of the borehole. Expansion shell anchored rockbolts have wide application in mining as well as in civil engineering. For application in permanent reinforcement systems, the void between the bolt and the borehole can be post grouted. With only one or two exceptions, mechanical anchors are designed for use in moderately hard to hard rock conditions. They are not recommended for use in *very* hard rock, since a very hard rock will prevent the expansion shell from "gripping" the rock, and the anchor will slip under load.

Mechanically Anchored Rockbolt — Expansion shell anchor

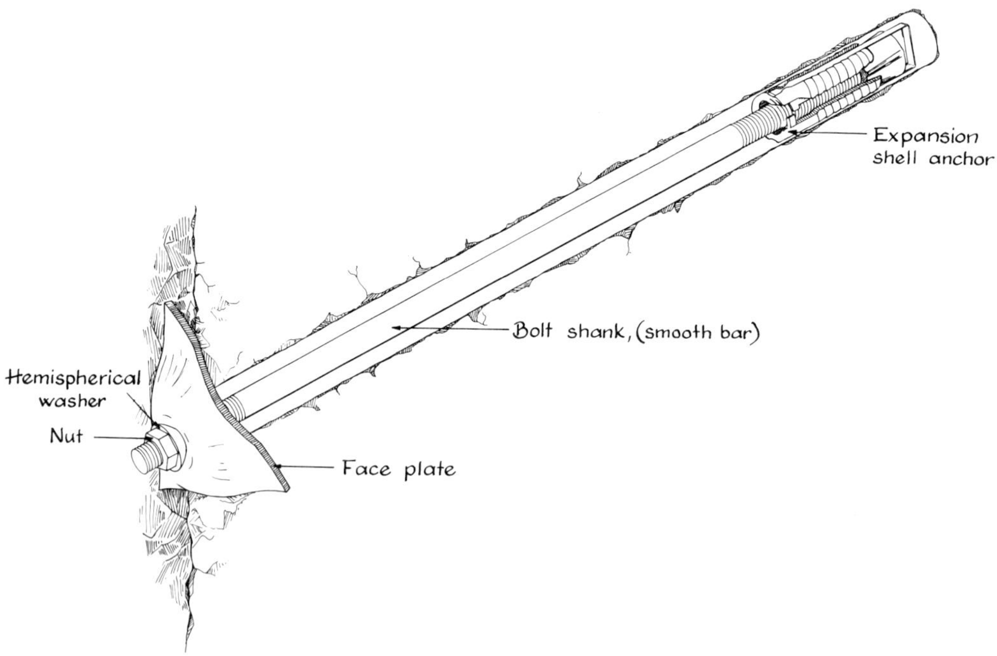

Typical technical data

Steel quality designation:	700 N/mm²	70 kp/mm²	102 kpsi
Steel diameter:	16 mm	16 mm	5/8 in
Yield load, steel:	140 kN	14 tons	15.5 tons(US)
Ultimate load, steel:	180 kN	18 tons	20 tons(US)
Ultimate axial strain, steel:	14 %	14 %	14 %
Weight of bolt without face plate and nut:	2 kg/m	2 kg/m	1.34 lb/ft
Bolt lengths:		any length required	
Recommended borehole diam.:	35—38 mm	35—38 mm (critical)	1 3/8 in

Advantages

Relatively inexpensive. The bolt gives immediate support action after installation. By rotating the bolt, a torque is applied to the bolt head and tension accumulates in the bolt. By post-grouting, the bolt can serve as permanent reinforcement. In hard rock, high bolt loads can be achieved. It is a versatile system for rock reinforcement, assuming hard rock conditions.

Disadvantages

Limited to use in moderately hard to hard rock. Difficult to install reliably, (see also section 4.1). Must be monitored and checked for proper tensioning. Loses bearing capacity as result of blast vibrations or when rock spalls off around borehole collar due to high rock stresses.

3.2 Grouted Rockbolts

Grouted rockbolts have been commonly used worldwide for the past forty years both in mining and civil engineering applications.

The most commonly used grouted rockbolt is the fully grouted rebar or threaded bar made of steel. Cement or resin are used as grouting agents. The rebar used with resin creates a system commonly used for tensioned rockbolts but the rebar or the threaded bar with cement grout can also be used for untensioned bolts. Both systems are used for temporary as well as permanent support under various rock conditions. The threadbar rockbolt is mainly used in civil engineering applications for permanent installation.

A few years ago it was predicted that the resin would, in general, replace the use of cement as grouting agent for fully grouted rockbolts. However, for a number of reasons, (mainly cost) this did not occur.

Grouted Rockbolt — Rebar

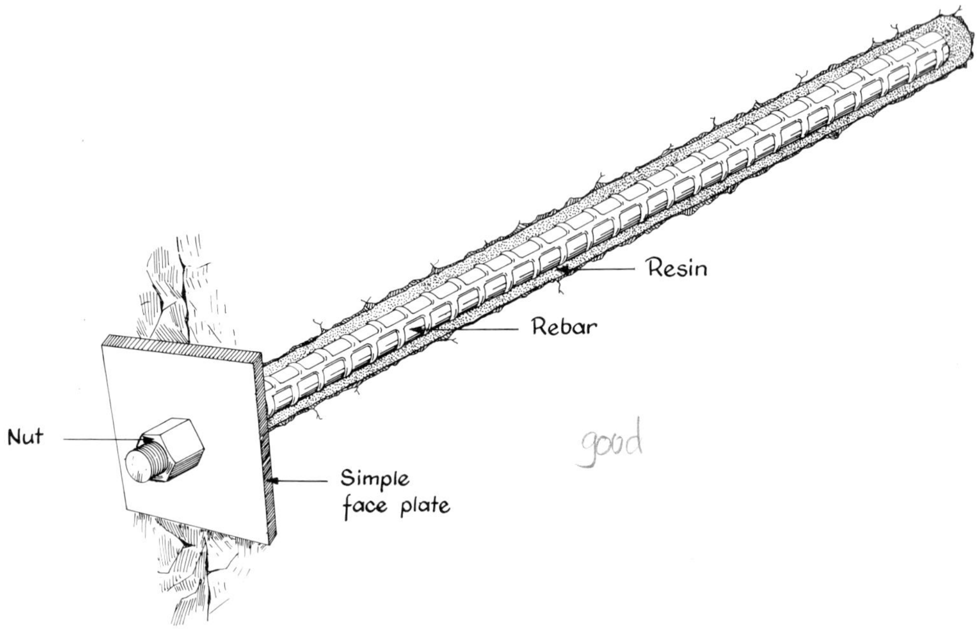

Typical technical data

Steel quality designation:	570 N/mm²	58 kp/mm²	83 kpsi
Steel diameter:	20 mm	20 mm	7/9 in
Yield load, steel:	120 kN	12 tons	13 tons(US)
Ultimate load, steel:	180 kN	18 tons	20 tons(US)
Ultimate axial strain, steel:	15 %	15 %	15 %
Weight of bolt without face plate and nut:	2.6 kg/m	2.6 kg/m	1.75 lb/ft
Bolt lengths:		any length required	
Recommended borehole diam.:	35±5 mm	35±5 mm	1 3/8 in

Advantages

The bolt gives rapid support action after installation. If a "fast-setting" resin is used for bottom anchoring of the bar, the fully grouted rockbolt can be tensioned, (see also section 4.2). High corrosion resistance in permanent installations.

Disadvantages

Difficulties with the resin cartridges in underground environment which can affect installation reliability, (see also section 4.2). Resin can be messy and hazardous to handle as well as wasteful. Resin has a limited shelf life.

Grouted Rockbolt — Dywidag® steel

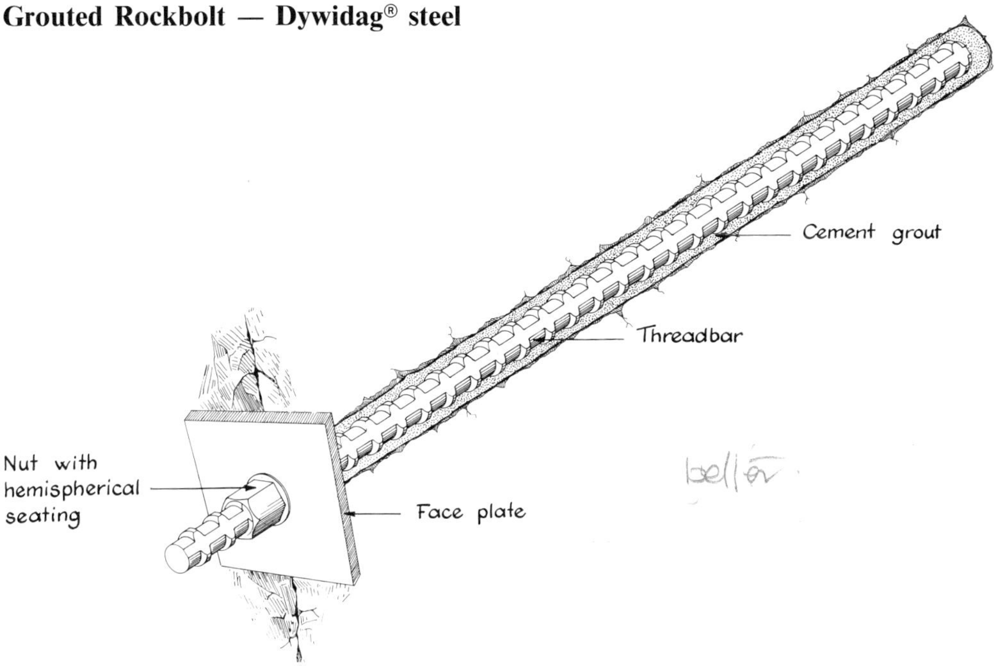

Typical technical data

Steel quality designation:	1080 N/mm²	110 kp/mm²	157 kpsi
Steel diameter:	20 mm	20 mm	7/9 in
Yield load, steel:	283 kN	28 tons	31 tons(US)
Ultimate load, steel:	339 kN	34 tons	37.5 tons(US)
Ultimate axial strain, steel:	9.5 %	9.5 %	9.5 %
Weight of bolt without face plate and nut:	2.6 kg/m	2.6 kg/m	1.75 lb/ft
Bolt lengths:	any length required		
Recommended borehole diam.:	35±3 mm	35±3 mm	1 3/8 in

Dywidag® is a registered trademark of Dyckerhoff & Widmann AG, F.R. Germany.

Advantages

Properly installed, it is a competent and durable reinforcement system. Affect of corrosive environments is minimal. The system gives high bolt loads in various rock conditions.

Disadvantages

Expensive. Tensioning of the rockbolt is possible only if special installation procedures are followed. Use of standard cement in the grout requires several days curing before the bolt can take load. Quality of grout is difficult to check and maintain constant.

3.3 Grouted Cable Bolts

Grouted cables have been used for reinforcement of structures in rock for the past twenty or thirty years. Cables were introduced to the mining industry, for the same purpose, over the last fifteen to twenty years.

The use of untensioned fully grouted cables in the mining industry is growing rapidly. Cable reinforcement is predominantly used in temporary support systems, STILLBORG (1985). Permanent reinforcement in mining has, to date, only been considered occasionally in relation to untensioned fully grouted cables. The use of cable reinforcement in a mining situation is, in general, quite different from applications in civil engineering where permanent, pre-stressed cables are considered. Pre-stressed cables for civil engineering applications is also the area where most development work has been conducted to date.

The most commonly used type of cable for cable reinforcement in the mining industry is a 15.2 mm 7-wire steel strand, normally installed in units of two. The cable is developed for pre-stressed concrete elements. The cable is often appropriately modified by the local supplier to improve its performance as a reinforcing element in rock.

The cable bolt described below is a new development. It is the first cable that has been designed for the purpose of rock reinforcement. As such, it displays some interesting properties for use both as a grouted rockbolt and as a traditional cable bolt for support of large volumes of rock. The cable has already gained considerable popularity within the Scandinavian mining industry.

There are a number of features in the use of a flexible cable instead of a rebar or a threadbar in a reinforcement system for grouted rockbolts. Some of these are: variation in bolt length is of no concern since the cable can be installed to any length from a reel even in narrow tunnels, and inexpensive, very high load bearing capacity, for mechanization variation in bolt length represents no problem.

Grouted Cable Bolt — Flexirope®

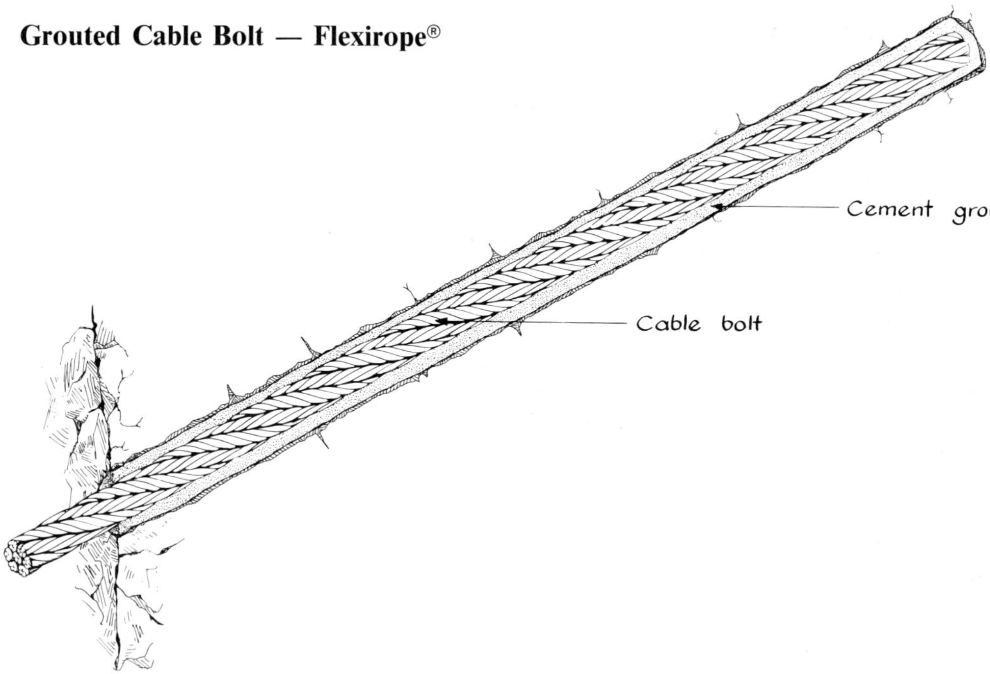

Typical technical data

Steel quality designation:	1770 N/mm²	180 kp/mm²	257 kpsi
Cable diameter:	28 mm	28 mm	11/10 in
Yield load, cable:	500 kN	50 tons	55 tons(US)
Ultimate load, cable:	500 kN	50 tons	55 tons(US)
Ultimate axial strain, cable:	3 %	3 %	3 %
Weight of cable:	3.1 kg/m	3.1 kg/m	2.08 lb/ft
Cable lengths:	any length required		
Recommended borehole diam.:	35 mm and above	35 mm and above	1 3/8 in and above

Flexirope® is a registered trademark of Scan Rope AB, Sweden.

Advantages

Inexpensive. Properly installed, it is a competent and durable reinforcement system. Can be installed to any length in narrow areas. The system gives very high bolt loads in various rock conditions, as well as high corrosion resistance in permanent installations.

Disadvantages

Tensioning of the cable bolt is possible only if a special installation procedure is adopted. The use of standard cement in the grout requires several days curing before the cable can be loaded.

3.4 Friction Anchored Rockbolts

Friction anchored rockbolts represent the most recent development in rock reinforcement techniques. Two friction anchored rockbolt types are available, the Split Set and the Swellex. For both types of rockbolt system, the frictional resistance to sliding of the rock on the steel (for the Swellex combined with mechanical interlock) is generated by a radial force against the borehole wall over the length of the bolt. Friction anchored rockbolts resemble mechanically anchored bolts in the sense that their installation and operation is unimpaired by wet rock conditions. For use in permanent installations, corrosion may, however, create a problem.

Although the two systems are described under a common heading, they display some major differences. These are related to the anchoring mechanism and their support action, as well as their installation procedure. Strictly speaking, only the Split Set is a true friction anchored rockbolt and as such it is sometimes called the "Split Set Friction Rock Stabilizer".

The anchoring mechanism of the Split Set rockbolt arises from frictional forces to a load which approaches the ultimate load bearing capacity of the bolt, when the bolt will slide. The bolt can then accommodate large displacements without failing.

The anchoring mechanism of the Swellex rockbolt depends on frictional forces combined with mechanical interlocking. The anchoring of the Swellex is provided by frictional forces to a load which approaches the ultimate load bearing capacity of the bolt. Mechanical interlock between the bolt and the rock then prevents the bolt from sliding. This property of the Swellex rockbolt implies that the full strength of the bolt is utilized. A higher pull-out resistance is then obtained.

The action of the Split Set rockbolt, to accommodate gross displacements by sliding is sometimes preferred, however. When required, this action can be obtained by a reduction in the inflation pressure of a standard Swellex. It has been shown that the Swellex can be adapted to a variety of ground conditions by changing the inflation pressure.

Both friction anchored bolt types are commonly used in the mining industry. However, their use in civil engineering applications is limited, but Swellex is increasingly used in tunnelling work.

Friction Anchored Rockbolt — Split Set®

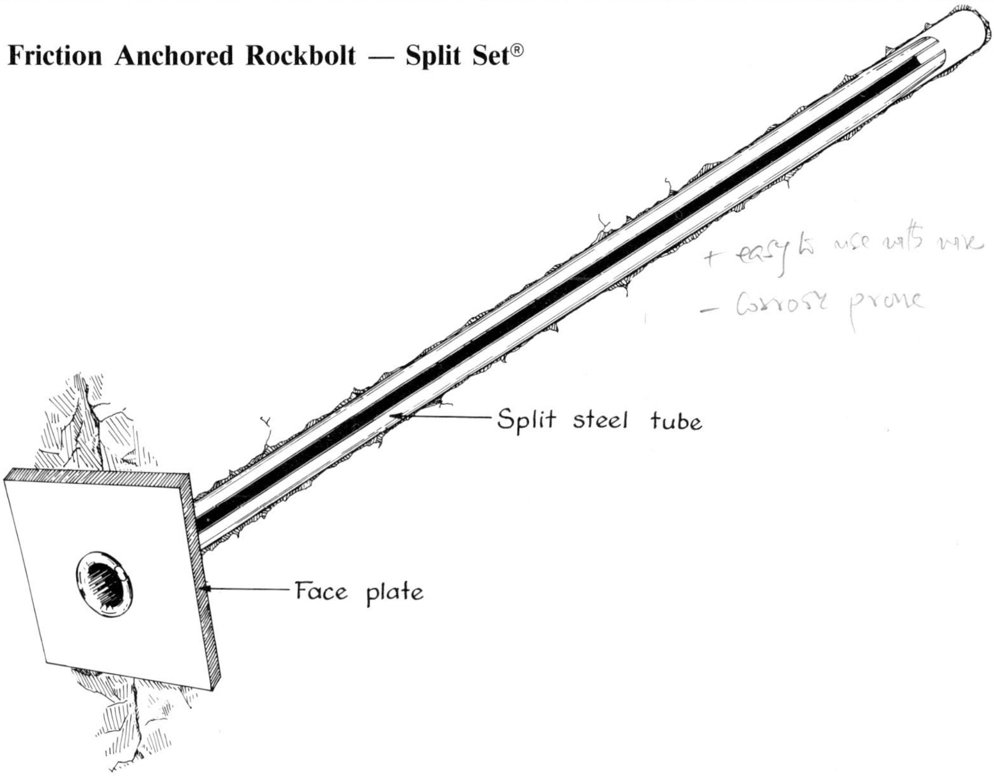

Typical technical data

Tube diameter:	39 mm	39 mm	11/2 in
Yield load, steel tube:	90 kN	9 tons	10 tons(US)
Ultimate load, steel tube:	110 kN	11 tons	12 tons(US)
Ultimate axial strain, steel tube:	16 %	16 %	16 %
Weight of bolt without face plate:	1.8 kg/m	1.8 kg/m	1.2 lb/ft
Bolt lengths:	0.9—3 m	0.9—3 m	3—10 ft
Recommended borehole diam.:	35—38 mm	35—38 mm (critical)	1 3/8 in

Split Set® is a registered trademark of Ingersoll-Rand Company, USA.

Advantages

Simple installation. Gives immediate support action after installation. No hardware other than a jackleg or jumbo boom for installation. Easy application of wire mesh.

Disadvantages

Relatively expensive. Borehole diameter is crucial in the prevention of failure during installation and in the provision of the intended holding force. Successful installation of longer bolts can be difficult. Cannot be used in long term installations unless protected against corrosion.

Friction Anchored Rockbolt — Swellex®

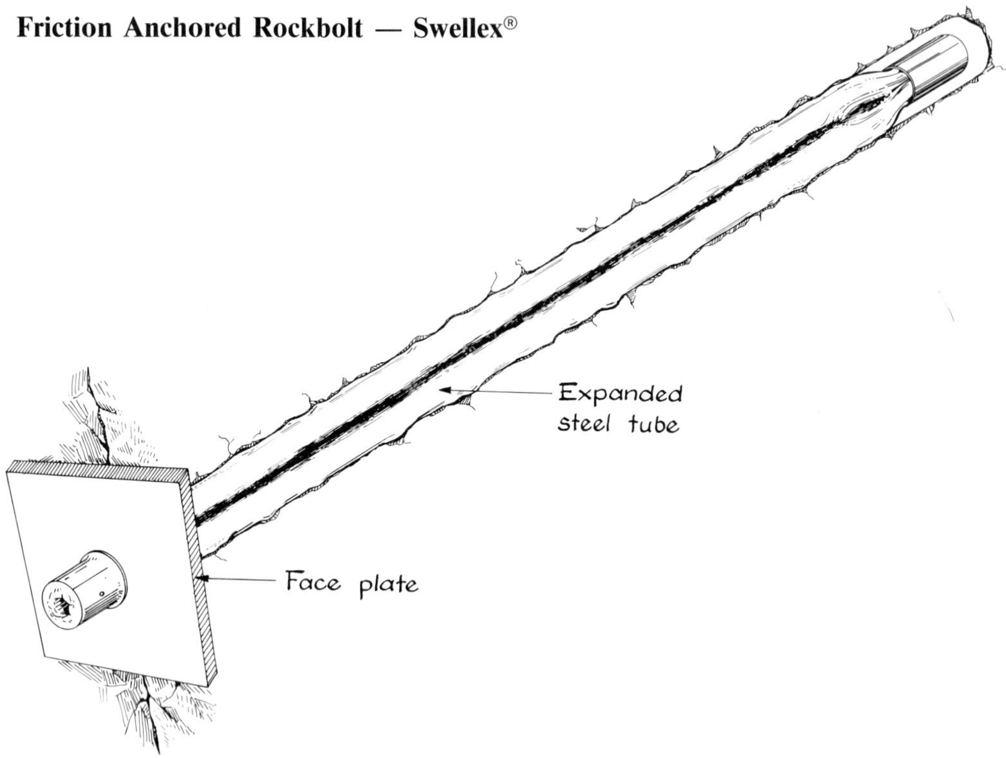

Typical technical data

Tube diameter:	26 mm	26 mm	1 in
Yield load, steel tube:	130 kN	13 tons	14 tons(US)
Ultimate load, steel tube:	130 kN	13 tons	14 tons(US)
Ultimate axial strain, steel tube:	10 %	10 %	10 %
Weight of bolt without face plate:	2 kg/m	2 kg/m	1.34 lb/ft
Bolt lengths:		any length required	
Recommended borehole diam.:	35±3 mm	35±3 mm	1 3/8 in

Swellex® is a registered trademark of Atlas Copco AB, Sweden.

Advantages

Rapid and simple installation. Gives immediate support action after installation. Can be used in a variety of ground conditions. The installation causes contraction in the bolt length. This effectively tensions the face plate against the rock surface.

Disadvantages

Relatively expensive. Corrosion protection required if used in long term installations. Requires a pump for installation.

3.5 Support Accessories to Rockbolt Systems

A number of support accessories are used together with rockbolts, apart from the face plate. Rockbolts are commonly used to attach different kinds of wire mesh and rock straps to the rock.

Face plates

A face plate is designed to distribute the load at the bolt head uniformly into the surrounding rock. To maintain the elasticity of the rockbolt system, the choice of face plate is crucial. Some commonly used face plates are shown in Fig. 2.

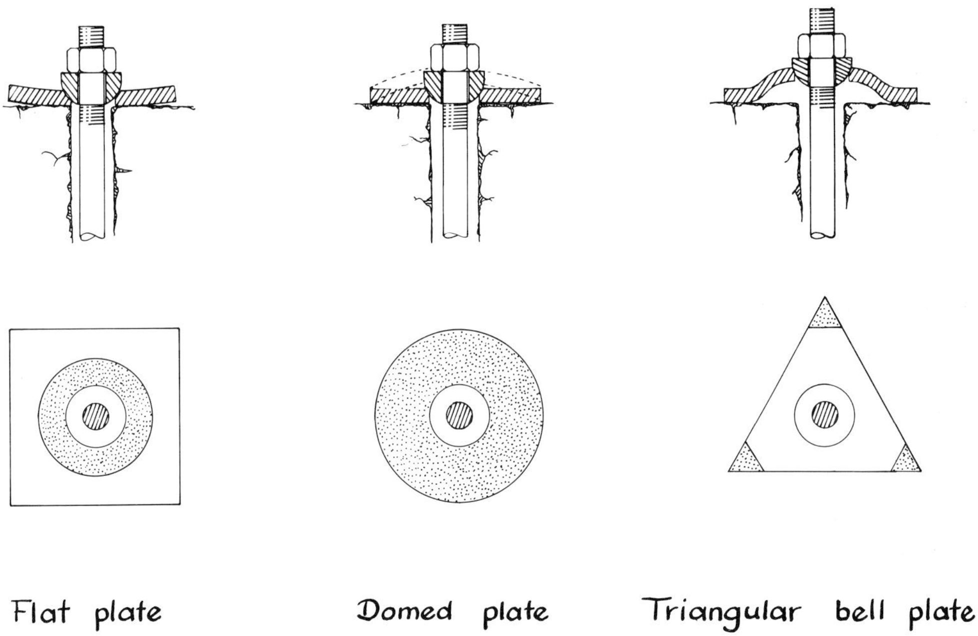

Fig. 2 Some commonly used face plates and their respective effective contact surface. Partly after SCHACH (1971).

The flat plate can be used when the rock surface is smooth and the bolt is installed perpendicular to the surface of the rock. If a hemispherical seating is added to the nut as in Fig. 2, the bolt can be installed obliquely to the rock surface without introducing unfavourable tensile stresses in the bolt. Another advantage with the hemispherical seating is that the nut will lie flat against its base. This permits more favourable tensioning of the bolt. The flat face plate, when loaded, is supported only at a few highly stressed points at the rock surface, (see Fig. 2).

Under sufficiently high pressure, the rock may crush at these points. The tension in the bolt will then be reduced. A setting of only one to two mm can reduce the tension in the bolt by 20—70%. The bolt must then be retensioned. This shortcoming in the flat face plate is greatly reduced with the domed or triangular bell plates, which have larger supporting areas. These plates also give the rockbolt system greater flexibility. If the bolt is first tensioned and then fully grouted the tension in the bolt is preserved and the bolt action is no longer dependent upon the face plate. When untensioned bolts are used, (e.g. fully cement grouted rebars), a simple face plate may be provided. This ensures an adequate face anchoring. It will also prevent debonding of the bar from the borehole at the rock face. Furthermore, face plates can give some support to the rock surface if the nut has been tightened with a spanner to ensure seating of the face plate. Many grouted rockbolts are installed without any face plate attached.

Wire mesh

Two wire mesh types are commonly used in combination with rockbolts. These are the chainlink mesh and the weldmesh.

The chainlink mesh is flexible and strong. It is usually used right up to the face, to prevent injury to personnel and damage to equipment from smaller pieces of rock or spalled flakes, see Fig. 3. The mesh should be attached to the rock at intervals of between 1 and 1.5 metres. Short grouted pins can be used between bolts. Depending upon the spacing between support points, a chainlink mesh can carry a considerable load of broken rock. The chainlink mesh can also be used in combination with wire rope lacing, see Part II of this handbook.

Weldmesh consists of steel wires set on a square or rectangular pattern and welded at each intersection. It is used for reinforcing shotcrete and is much more rigid than the chainlink mesh. The latter is unsuitable for shotcrete reinforcement. Generally, the weldmesh is attached to the rock by means of a second plate washer and nut placed on existing rockbolts, alternatively pinned to the rock with additional short bolts and plate washers. Typically, 4.2 mm wires set at 100 mm intervals are used for reinforcing shotcrete.

Shotcrete

Shotcrete is commonly used in combination with rockbolts, in a wide range of ground conditions, particularly in underground civil engineering applications. Also in underground mining, shotcrete is increasingly used to provide support to the rock surface between rockbolts.

There are two basic types of shotcrete, the dry-mix shotcrete where the mix is dry and the water is added at the nozzle. The wet-mix shotcrete has basically the same components as the dry-mix, except that the water is already added in the mixer. In the case of the wet-mix, any accelerator must be added at the nozzle. The dry-mix shotcrete type is most widely used.

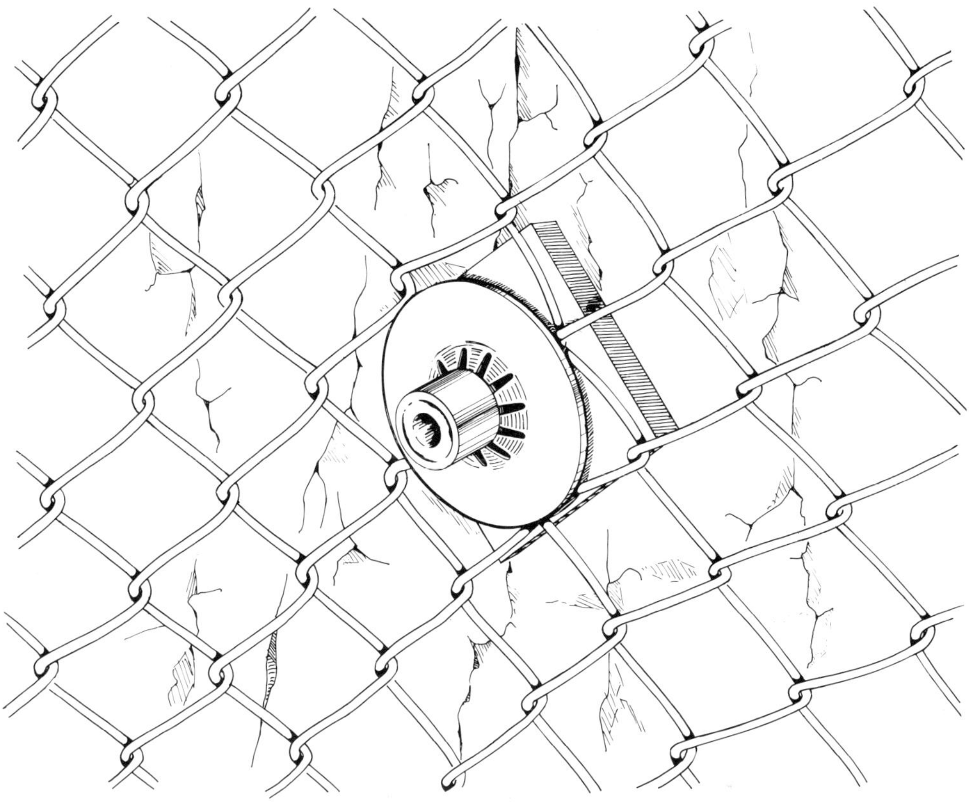

Fig. 3 Chainlink mesh anchored by rockbolts and used as protection from loose rock between bolts.

The quality of the shotcrete placed, depends upon the materials used and the mix design. However, this, together with the method of placement, is also very much a matter of skill and judgement among the operators.

A typical mix contains the following percentage of dry components by weight:

Cement	15—20 %
Coarse aggregate	30—40 %
Fine aggregate or sand	40—50 %
Accelerator	2— 5 %

The water/cement ratio (by weight) for dry-mix shotcrete lies in the range 0.3—0.5 and is adjusted by operator to suit local conditions. The addition of 50 mm long and 0.4—0.8 mm diameter steel fibres has been found to improve the toughness, durability, and shear and flexural strengths of shotcrete, and to reduce the formation of shrinkage cracks. Steel fibres are added according to the specifications of the manufacturer.

In applying the shotcrete, the nozzle should be kept as nearly perpendicular to the rock surface as possible, and at a constant distance of about 1 m. A permanent shotcrete lining is usually between 100 mm and 500 mm thick, the larger thicknesses being placed in a number of layers. Details of reinforced shotcrete application in combination with rockbolt reinforcement is illustrated in Fig. 4.

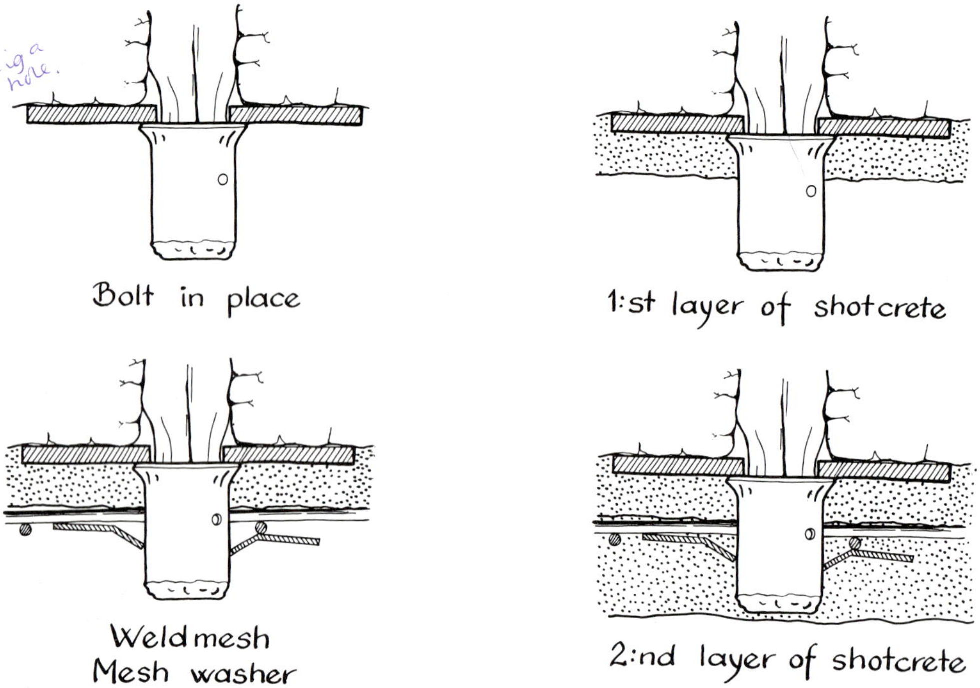

Fig. 4 Details of reinforced shotcrete application in combination with rockbolt reinforcement.

Rock straps

Rock straps are commonly made of 6 mm (1/4 in) steel which is approx. 100 mm (4 in) wide and of variable length. Straps are used where rock conditions are poor and a ravelling failure around the collars of the rockbolts has been experienced. To inhibit this type of failure, the rockbolts can be connected together using steel straps supporting the surface of the rock.

Chapter 4

Rockbolt Installation

The installation of rockbolts and cable bolts should be organized in such a way that the installation becomes an integral part of the excavation cycle and gives temporary support to the rock. Since many types of rockbolts can be used in permanent reinforcement systems, it is important to consider the "temporary" reinforcement system as a part of a later, additionally installed, long term support system. The type of rockbolt should be chosen with the complete ground support system in mind. The installation should be performed by qualified, properly trained and experienced personnel.

In general, rockbolt and cable bolt installation is a problem of working accessibility. A manual procedure is still therefore the most common installation technique. Today, however, fully mechanized bolting rigs are available for all types of rockbolts and cable bolts. These rigs are capable of performing the complete installation cycle. The bolting capacity of a fully mechanized rig is often large, and the quality of installation obtained is generally better than that obtained from manual installation. Also, from safety and health aspects, the fully mechanized bolting rig provides very good working conditions for the operator. It is necessary to have a bolting programme which is sufficiently extensive to justify the cost of an elaborate rig.

Many different varieties of rockbolts and cable bolts are in use. Numerous different installation techniques also exist. These are frequently developed to meet local requirements and limitations. Before reviewing installation procedure, the importance of proper scaling before rock reinforcement must be emphasized. One of the common causes of accidents in underground excavations, involving falling rock, is inadequate scaling after a blast. Mechanical scaling is always to be preferred, since it generally gives a better result and is safer for the scaling crew. Where mechanical scaling is not possible, manual scaling should be performed so that smaller blocks do not become detatched as a result of deformation of the excavation, subsequent blasting or drilling in an adjacent area.

4.1 Installation of Mechanically Anchored Rockbolts

Borehole diameter is critical for the installation of expansion shell anchored rockbolts. The bolt cannot be installed in an undersized or oversized hole. To install, the bolt is pushed into the hole until the face plate is in contact with the rock. The nut on the bolt is then rotated until a preset torque is reached. This can be accomplished by the use of a tool which automatically shuts off, at a pressure corresponding to the preset torque. The recommended applied torques range from 135 to 340 Nm (100-250 ft-lb) or 4.5 kN (1000 lb) load above or below the 50% limit of the bolt yield load or anchorage capacity whichever is lower, PENG (1978). As the torque is applied to the bolt, a tension develops in the bolt, provided the expansion shell firmly grips against the borehole wall. The torque — tension relationship can vary considerably for a specific applied torque. Nevertheless, a linear relationship between bolt tension and torque exists, and this can be written:

$$P = CT \tag{1}$$

where P is the bolt tension in N (pounds) and T is the applied torque in Nm (foot-pounds). C is the proportionality constant between the bolt tension and the applied torque. As a "rule of thumb", C is equal to 50 for a 16 mm (5/8 in) bolt and 40 for a 19 mm (3/4 in) bolt, PENG (1978). C should be given in the appropriate units. Factors that affect the value of the constant C are the applied upward thrust on the bolt during installation and the angle of installation.

The effectiveness of an expansion shell anchored rockbolt depends mainly on the grip of the shell against the borehole wall. Apart from the borehole size, the rocktype and the integrity of the rock at the zone of anchorage will affect the gripping force. In hard rock conditions high bolt loads can be achieved. However, if the bolt is placed within 10—20 m (30—60 ft) of a blast, the bolt can lose its tension and must be retensioned. In weaker rocks, the effectiveness of the bolt is reduced by local crushing of the rock by the ribbed sleeves of the shell. Clay-filled joints can, in otherwise good conditions, cause serious problems. In very weak rocks like fractured shale or mudstone, the use of mechanical rockbolts in general, is not recommended.

The installation and the effectiveness of the bolt is independent of the water conditions in the borehole when the bolt is used for temporary support. Post-grouting is performed when the bolt forms part of a permanent reinforcement system.

Installation time, for an expansion shell anchored rockbolt of 2 m (6.5 ft) length, (excluding the time for borehole drilling) is typically 75 sec.

Fig. 5 Tensioning of an expansion shell anchored rockbolt using a torque wrench.

4.2 Installation of Grouted Rockbolts

The installation procedures used for grouted rockbolts are very similar, whether cement or resin grout is used. The grout can be placed in the borehole either by pumping or by using cartridges. When the grout is pumped into the borehole, common practice is to push the grout tube into the bottom of the hole and fill the hole as the tube is retrieved. It is important to ensure that the tube is in full contact with the cement/resin, as it is retrieved, to prevent the formation of airpockets. When the hole is filled, the bolt is pushed into the hole through the grout, until it hits the bottom of the hole or full contact is made between the face plate and the rock surface.

To keep the grout in vertical upholes the grout must be sufficiently thick. A cement grout with a water/cement-ratio (by weight) of between 0.30 and 0.35 is commonly used. The cement grout can be used with an additive that will accelerate the setting of the grout like quick-setting resins are available. If cement mortar is to be used, ready-mixed mortar is recommended.

When the grout is pumped into the hole, there is a problem of sealing the collar of the borehole. Proper sealing is important, especially in vertical upholes. In upholes, the sealing prevents excessive grout from leaving the hole after bolt installation, and it also stops the bolt dropping out of the hole until the grout has set.

When the grout is pumped into the borehole, the borehole size is not critical for the final grouting result.

A more convenient and quicker method of grouted rockbolt installation is to use cartridges. Both cement and resin cartridges are available. The most commonly used cement cartridge contains cement and additives within a porous container which soaks in water, before it is placed in the borehole (like the Cembolt product).

Mixing is required for the resin cartridge. This is not the case with the cement cartridge. The mixing of the resin cartridge takes place when the bolt is rotated through the cartridge. Cement grout cartridges can also be used in combination with resin cartridges to reduce the cost of fully resin grouted rockbolts, see Fig. 6.

The problem of excessive grout leaving the borehole is not usually encountered when cartridges are used, since the cartridge normally produces a thicker mix. The problem of keeping the bolts in upholes until the grout has set remains. The use of a simple sealing at the hole collar provides the best solution to this problem.

When cartridges are used to place the grout in the borehole, the hole size becomes critical both for proper mixing and filling of the hole. Manufacturer specifications should be followed.

Bolt tensioning can be applied either by a tool which rotates the nut on the bolt until a preset torque is reached, or by direct pull. A torque wrench or a machine which is set to stall at a pressure corresponding to the preset torque is usually adequate. The problem with this method, applied to grouted bolts, is almost identical to that discussed in section 4.1 for mechanically anchored rockbolts. If the bolt is to be tensioned at higher loads (exceeding approximately 100 kN (10 tons)), tensioning in direct pull with a hydraulic bolt tensioner is to be recommended, HOEK and BROWN (1980), Fig. 6. The tensioner should have an easily read calibrated gauge for the load. Provided the capacity is sufficiently large, the hydraulic bolt tensioner can also be used for pull-out testing of the bolts.

Using the technique described above, for pumping the grout into the hole, the bolt cannot be tensioned unless it is partly debonded. Debonding of the bolt top section can be performed by using PVC tape or a high density polyethylene sleeve. By using a quick-setting resin cartridge over the bond length and slow-setting resin cartridges over the free length, the bolt can be tensioned as soon as the former has gained sufficient strength. Typically, a resin grouted bolt can be tensioned within 1 to 5 minutes after installation, while a cement grouted bolt requires a couple of hours to set, depending on the additives used and rock temperature.

The fully grouted rockbolt is one of the most common of all rockbolt systems. It is perhaps also the most versatile system. When cement grout is used in combination with a standard rebar, this is often the cheapest type of rock reinforcement. It is, however, important to check the quality of both the resin and the cement before use, since both agents are products which are sensitive to storage underground for extended lengths of time. Some results from tested grouted rockbolts, which indicate the difficulty of obtaining a good grouting of the bolt, are given in Chapter 5.

Installation time, for a fully resin grouted rockbolt of 2 m (6.5 ft) length, (excluding the borehole drilling and the curing time) is typically 75 sec.

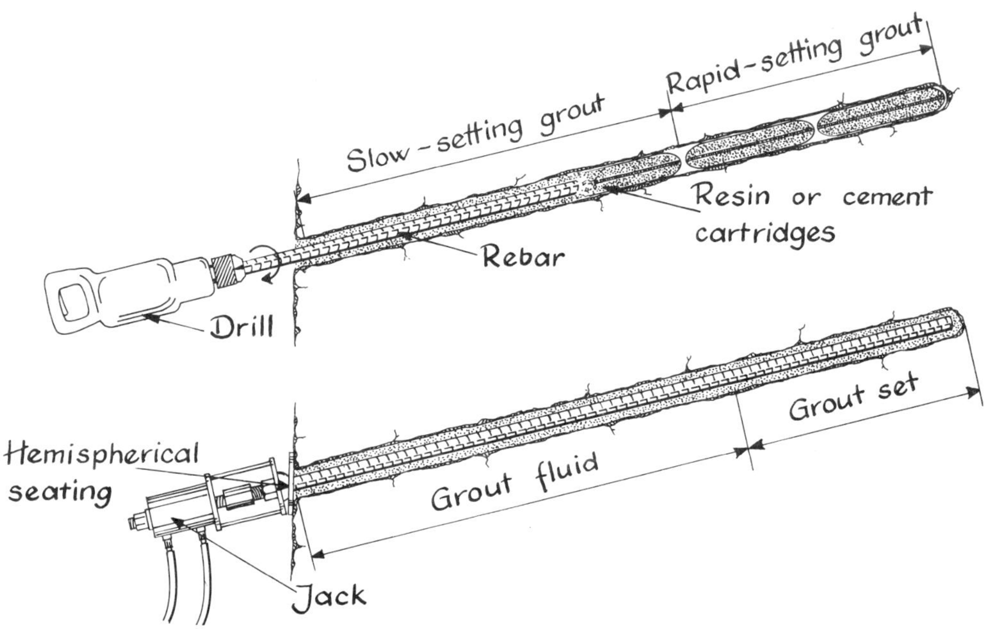

Fig. 6 Principle of installation and tensioning of a fully grouted rockbolt using an hydraulic bolt tensioner.

4.3 Installation of Grouted Cable Bolts

Short cables, (cable length less than 6 metres or 20 ft) can replace the rebar or the threaded bar in a reinforcement system. The installation procedure used for short cables is the same that is used for the grouted rockbolts, except that resin is rarely used as the grouting agent.

Installation time, for a fully cement grouted cable bolt of 2 m (6.5 ft) length (excluding the borehole drilling and the curing time) is typically 100 sec.

For grouting long cables, (longer than 6 metres), cement is exclusively used as the grouting agent. The installation procedure for long cables in horizontal and downwards inclined boreholes is as follows. The grout hose is inserted to the bottom of the borehole and retrieved as the hole is filled with cement. After the hole has been filled, the cable is pushed down the hole until it reaches the bottom of the hole. It is recommended that the inserted length of cable is measured accurately and compared with the drillers borehole log-sheet. This procedure will prevent the cable from not being fully inserted. To facilitate the installation of long cable bolts, a pointer made of steel or plastic is often attached to the end of the cable, see Fig. 7.

Installation of long cables in horizontal to upwards inclined boreholes is illustrated in Fig. 7. First, the cable is inserted to the bottom of the borehole, together with a stiff nylon or polyethylene breather tube. After the grout tube has been inserted some 0.5 metres (1.5 ft) into the hole, the hole is sealed off at the collar. In steep upholes, it is necessary to lock the cable in place to prevent it from sliding out of the hole. The cable is locked by a top anchor arrangement or by some locking arrangement at the collar of the hole. Once the hole is sealed off, and the cable locked in place, the hole is filled with cement grout and the air trapped inside the hole escapes through the breather tube. The hole is filled with cement grout, when weak grout flows out of the breather tube. Finally, the grout tube is sealed and cut as illustrated in Fig. 7.

Results by STILLBORG (1984) show that even though the bond strength between the cable and the grout is improved when the water/cement-ratio decreases, the choice of water/cement-ratio is limited by the capacity of available mixing and pumping equipment for the grouting of long boreholes. It was concluded that an 0.4 water/cement-ratio (by weight) is the most suitable mix for practical use in the installation of long cables.

The bond strength is improved by using additives in the cement grout. However, complex working practices give final grouting results which are difficult to control. This is particularly true in a mining environment. In conclusion, the use of any additive, including sand, will make the installation process of the cables or rockbolts more costly and difficult to control. Cement cartridges are never used for installation of long cable bolts.

Mechanical equipment for installation of cables in any direction of the borehole is commercially available, see Part II of this handbook. Cable bolting using long cables in steep upholes, can be very difficult and inefficient, without the assistance of a specially designed cable bolting rig.

Installation time, for a fully cement grouted cable bolt of 20 m (65 ft) length (excluding the borehole drilling and the curing time), but with the assistance of a mechanical cable feeder, is, in a downward inclined borehole, typically of the order of 30 min.

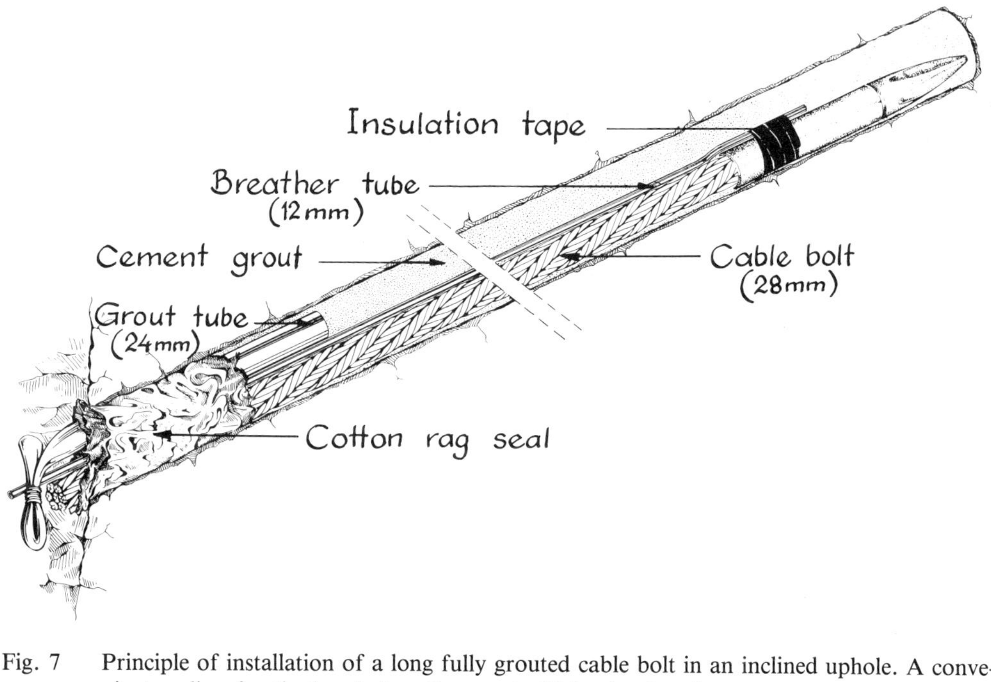

Fig. 7 Principle of installation of a long fully grouted cable bolt in an inclined uphole. A convenient sealing for the borehole collar must still be developed.

4.4 Installation of Friction Anchored Rockbolts

The installation procedure for the two friction anchored rockbolts, the Split Set and the Swellex, are quite different.

The Split Set rockbolt is forced into a borehole. The borehole has a slightly smaller diameter than the bolt. A stoper or a jackleg drill is often used. Here the installation is sometimes a problem in confined areas. An incorrect borehole diameter is the major cause of failures during installation. The borehole diameter is also of critical importance to the support action of the Split Set after installation. If the hole diameter is too small, the bolt becomes difficult to install. If the hole diameter is too large, the holding force of the bolt is not sufficient. The bolt cannot be tensioned but it gives immediate support after installation. It is not usually suitable for permanent reinforcement. The bolt can be used in a variety of rock conditions but is less suitable in rocks which are very fractured and/or soft.

Installation time, for a Split Set rockbolt of 1.8 m (6 ft) length (excluding the borehole drilling time) is typically 40 sec.

The installation of the Swellex rockbolt is simple. It does not require the help of any mining machinery once the borehole is drilled. In manual rock bolting, the bolt is inserted in the hole and a high pressure water pump is used to expand the bolt within the borehole. The pump is carried manually. Inflation forces the steel into the irregularities of the borehole and thus provides frictional resistance, but the expansion of the bolt also causes a contraction in the length which effectively tensions the face plate against the rock surface. The bolt provides full support action immediately after installation and borehole diameter is not critical. A correct installation procedure of the bolt does not rely on operator judgement since the pressure of the pump is set at the specified level by the operator and this is his only input to the system. The pump stops automatically at the pre-set pressure, indicating that the bolt is properly installed.

The bolt can be adapted to a wide variety of ground conditions by adjusting the standard inflation pressure of 30 MPa (300 bar). In ground conditions where large rock movements occur, it has been shown that by reducing the inflation pressure to 24 MPa (240 bar) the bond strength was reduced and the Swellex was allowed to slide, Morland and Thompson (1985). In the opinion of the author, this is a procedure which is only recommended when the design of the support system is based on results of pull-out tests (the relationship between the bond strength and the inflation pressure has been established) and a detailed knowledge of rock mass response upon unloading.

The Swellex should not be used in long term installations unless it is protected against corrosion.

Fig. 8 Installation of a Split Set rockbolt.

Both types of friction anchored rockbolts can be used in water bearing rocks, even under high water pressure.

Installation time, for a Swellex rockbolt of 2.4 m (8 ft) length (excluding the borehole drilling time) is typically 25 sec.

Fig. 9 Installation of a Swellex rockbolt.

Chapter 5

Testing of Rockbolts

Failures of rock reinforcement systems can occasionally be attributed to the material properties of the bolt itself. The number of such cases are, however, negligible compared with those which relate to the installation of the rockbolts. Common causes of rock reinforcement failure are:

Incorrect choice of rockbolt system for a given application
Inappropriate setting of borehole
Borehole too long or too short
Incorrect borehole diameter
Borehole not properly flushed
Inadequate grout placement
Inadequate bond length
Dirty steel
Grouting agents out-of-date
Inappropriate grout mixture
Damage to breather tube
Inadequate sealing of borehole
Inadequate lubrication of nuts, washers, and threads
Incorrect installation procedure
Inadequate test programme
No monitoring of reinforcement system

The number of causes which can seriously affect the performance of the rockbolt reinforcement system are almost unlimited. It is therefore important that the work is performed by trained personnel. Even so, the testing of the rockbolt system is important to gain confidence in the final result. There are a number of testing methods available to control the quality of the installed bolt. A number of examples of testing methods which can be applied to the rockbolts described in this handbook are shown below.

Type of rockbolt	*Method of testing*
Expansion shell anchored rockbolt:	Pull-out test, torque wrench, load cell.
Grouted steel bars:	Pull-out test, overcoring and electronic non-destructive testing.
Grouted steel cables:	Pull-out test and overcoring.
Friction anchored rockbolts:	Pull-out test, overcoring, indirect non-destructive testing.

Pull-out test

A pull-out test is similar to the tensioning procedure, except that the bolt is now loaded to failure, see the hydraulic jack system shown in Fig. 6. Pull-out force and displacement are recorded.

Torque wrench

A torque wrench can be used to measure indirectly the load on the bolt. A torque is applied until the nut starts to turn. The measured torque can then be related to the load, see Fig. 5.

Load cell

The load on the bolt can be measured by a load cell which is placed between the face plate and a bearing plate at the surface of the rock. The load cell will, when installed correctly, give a very accurate reading of the load on the bolt as a function of time.

Overcoring

Overcoring by means of diamond drilling is used either to determine the integrity of the grout or to check corrosion.

Indirect non-destructive test

The Swellex rockbolt can be checked for proper installation as well as damage with the use of the waterpump that is employed to expand the bolt. A properly installed and faultless bolt will be confirmed, at installation, by the pump automatically stopping at the pre-set pressure. In the later life of the bolt the pump can again be used in the same manner, to check for damage or failure of the bolt. This can be done repeatedly, since it is a non-destructive test of the bolt.

Electronic non-destructive test

For cement and resin grouted steel bars the quality of the grouting as well as the bolt length can be checked by an electronic non-destructive testing device based on the principle of reflected stress waves. A specially designed sensor containing piezoelectric crystals is pressed against the free planar end surface of the bolt. Compressional and flexural elastic waves are transmitted into the bolt. When the waves travel along the bolt, some energy is transferred through the grout into the rock and thus the wave amplitude decreases. The elastic waves are reflected from the inner end of the bolt. The reflected waves are recorded at the outer end of the bolt by the piezoelectric crystals. If the grout surrounds the bolt fully, and is of good quality, the amplitude of the reflected wave is more completely damped than if the grout is of poorer quality. From the time interval between the excited and reflected wave, it is possible to calculate the length of the bolt. The amplitude of the reflected wave, or sometimes those of successive reflections, can be analyzed, and the probable condition of the rockbolt can be estimated on the basis of calibration tests.

The importance of testing of the rockbolt system after installation has been demonstrated by Bergman et al. (1983), for example.

Fully cement grouted rebars installed during 1975 and 1980 in an underground mine were tested by using the above mentioned principle of electronic non-destructive testing technique. The instrument, (called a Boltometer), gave the following class distribution for the quality of the grouting of the bolts;

A 35% — B 25% — C 25% — D 15%

where the tentative Boltometer index for cement grouted bolts is as follows;

Class	Boltometer reflection	Estimated bolt performance
A	No reflection	Optimum (up to 2.3 m bolt length)*
B	One small reflection	Reduced
C	One large or several minor reflections	Insufficient
D	Many large reflections	Very poor or non-existent**

* If the bolt is longer than 2.3 m and perfectly grouted in the outer 2.3 m part of the borehole, no conclusion can be drawn from the grouting quality of the innermost part.

** The bolt can (probably) be pulled out of the borehole by a hydraulic jack.

Despite the fact that the Boltometer instrument still requires further refinement and development, it does give a reliable indication of the quality of the grouting. The above results reflect a large variation in the quality of grouting of the bolts tested. This may well affect the load bearing capacity of the bolts as well as the system of reinforcement.

5.1 Load Bearing Capacity of Rockbolts

The load bearing capacity of a rockbolt or a cable bolt depends essentially on the type of bolt. For the same bolt type the load bearing capacity can, however, vary considerably. Examples of the causes of this variation can be found among the causes of rock reinforcement failure listed in the beginning of this chapter. The test results on grouted rockbolts given above, illustrate the importance of a carefully controlled installation procedure to avoid possible large variations in the load bearing capacity of the bolts.

For rock reinforcement design purpose, pull-out test results from the rock formation considered are of interest. Pull-out test data of correctly installed bolts provides a measure of the load bearing capacity as well as a basis for comparison of different

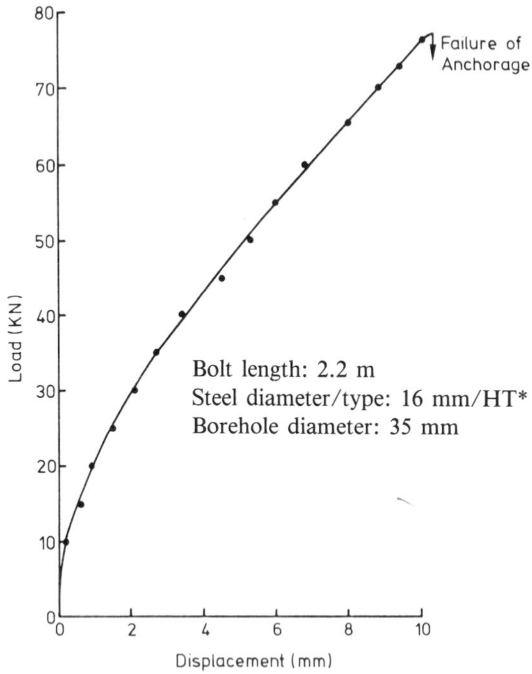

Fig. 10A Mechanically anchored rockbolt.

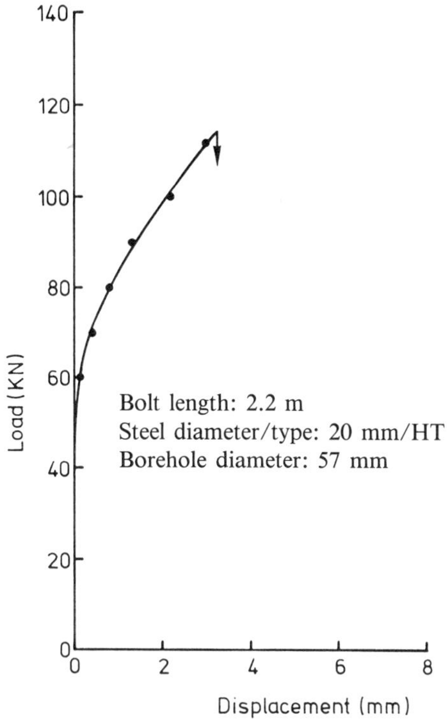

Fig. 10B Cement grouted rebar.

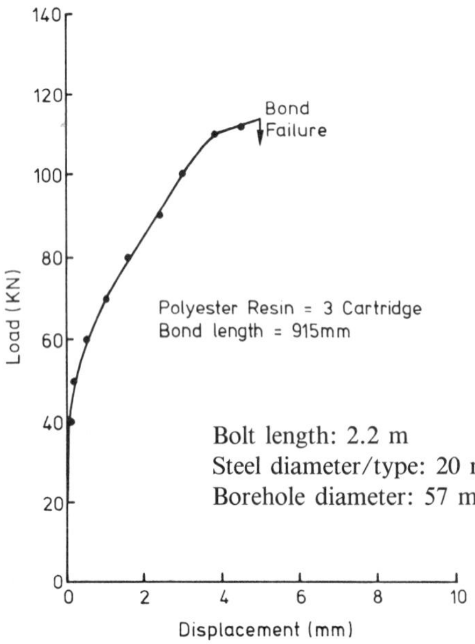

Fig. 10C Resin grouted rebar.

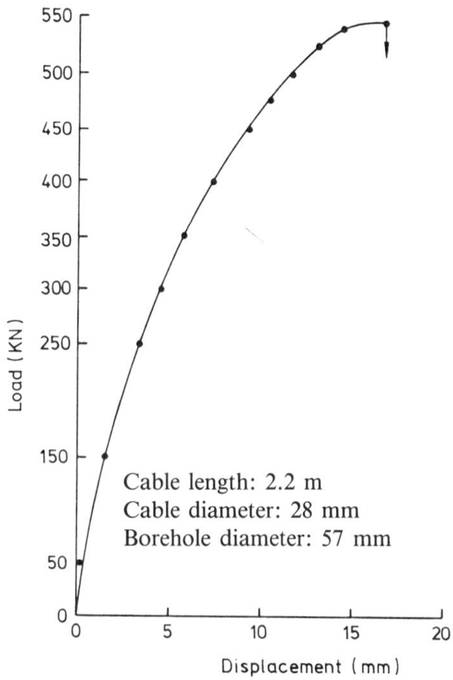

Fig. 10D Cement grouted Flexirope.

* HT = High tensile steel

Note the different scales of the diagrams!

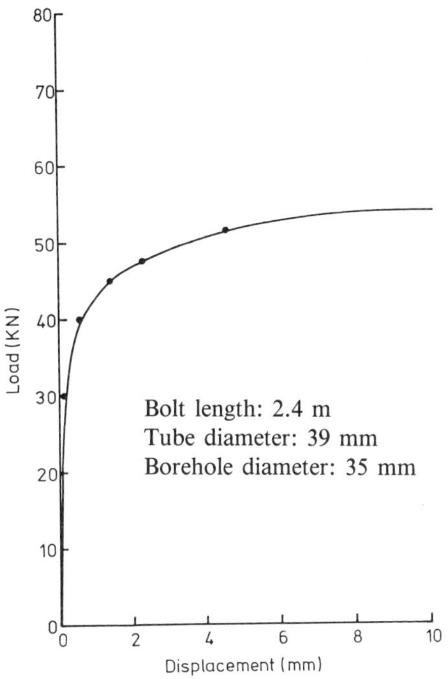

Fig. 10E Split Set.

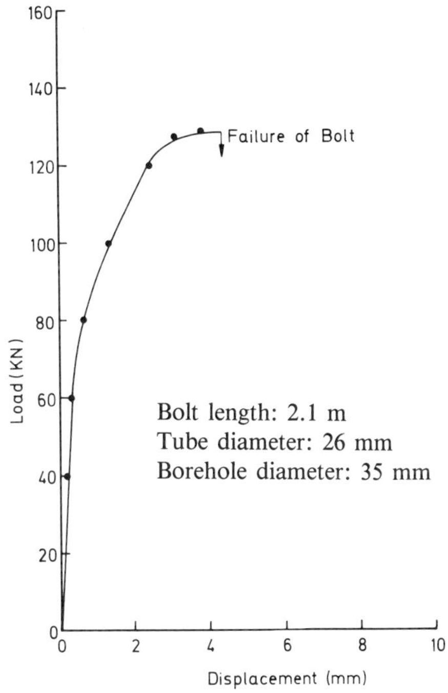

Fig. 10F Swellex.

Note the different scales of the diagrams!

bolt types. This does not imply that pull-out tests should be used exclusively in the evaluation of the appropriate rock bolting system for a given situation. The results obtained should only serve as a guide.

Above, as an example, a series of pull-out test results are given for mechanical, cement and resin grouted bolts, the Flexirope cable bolt, the Split Set and the Swellex friction anchored rockbolts. The tests were all conducted in the same massive dolomite rock formation, Singh (1985) except for the Flexirope, which was tested in a massive quartz-bearing porphyry, Stillborg (1986).

It is important to recognize the limitation in generality of the above test results. It is also necessary to appreciate the fact that any pull-out test results are specific to the ground conditions in which they were obtained as well as the installation procedure adopted. Caution should therefore be exercised when the pull-out test results are considered on an individual basis. However, because the pull-out test results of the rockbolts have been obtained from the same ground conditions, the results can be used as an index for comparison. The pull-out test result of the Flexirope should not, without reservations, be compared with the remaining results since it is obtained from a different rock type.

If a rockbolt is subjected to some component of shear, the load bearing capacity of the bolt is reduced, compared with a bolt which has been loaded in pure tension. The reduction in strength depends upon the bolt type and the angle between the bolt and the joint surface. Friction anchored rockbolts are less suitable in their ability to resist shear displacements than fully cement or resin grouted bolts. Their reduction in strength when they are sheared is larger than that obtained with solid steel bolts, LUDVIG (1984).

When the angle between the bolt and the joint surface is less than 40° the failure of the bolt is in tension. If the angle is greater than 40° the failure is in combined tension and shear. As the angle between the bolt and the joint surface approaches 90° the failure approaches that of a shear type of failure, BJURSTRÖM (1973).

Shear test results on cables have shown that a cable can sustain a much larger shear displacement than a solid steel bolt. This is due to the basic difference in construction of the two reinforcing members. The wires of the cable can redistribute themselves as the cable is sheared. This is particularly true for a multi-wire cable like the Flexirope, STILLBORG (1984).

Chapter 6

Design Considerations

The design of rockbolt systems is basically a matter of being able to predict the magnitude, directions and types of stresses which arise from the virgin stresses and their influence on the rock mass surrounding an excavation. In practice, this is not a trivial problem since stress changes in a rock mass depend on a number of factors such as excavation geometry, rock mass conditions, inherent rock stress situation etc. In addition, the effect of the reinforcement on the rock mass and the anticipated life time of the reinforcement must be considered. The following list gives some of the factors that influence the design of a ground support system;

Purpose of excavation	**General stress state**
Temporary or permanent reinforcement	**Rock mass conditions**
Safety requirements	**Groundwater conditions**
Practical considerations	**Possible failure modes**
Excavation geometry	**Support pressure required**
Available access	**Time for support action**
Extraction technique	Previous experience
Proximity of adjacent excavations	**Cost and availability**

In this chapter some of the most important of the above design considerations for rockbolt systems will be reviewed and whenever possible, examples will be given on how these considerations can be quantified.

6.1 Rock Mass Conditions

A geologist divides rocks into three different classes, depending upon their origin;

— Sedimentary rocks
— Metamorphic rocks
— Igneous rocks

Sedimentary rocks consist of an assemblage of detrital particles and possibly pebbles from other rocks in a matrix of materials such as clay minerals, calcite, quartz, etc. From their nature, sedimentary rocks contain voids or empty spaces, some of which may form an interconnected system of pores. Metamorphic rocks are produced by the action of heat, stress, or heated fluids on other rocks, sedimentary or igneous. Igneous rocks consist of a completely crystalline assemblage of minerals such as quartz, plagioclase, pyroxene, mica, etc. Typical mechanical properties of some common rock types are listed in Table 1.

Table 1 Typical Static Mechanical Properties of Some Common Rock Types*

Rock Class	Rock Type	Unconfined Compressive Strength σ_c [MPa]	Tensile Strength σ_t [MPa]	Modulus of Elasticity E [GPa]	Point Load Index $I_{s(50)}$ [MPa]	Angle of Friction ϕ-degrees
	Limestone	50—200	5—20	20—70	0.5—7	33—40
Sedimentary	Mudstone	5—15	—	—	0.1—6	—
rocks	Sandstone	50—150	5—15	15—50	0.2—7	25—35
	Siltstone	5—200	2—20	20—50	6—10	27—31
	Shale	50—100	2—10	5—30	—	27
Metamorphic	Gneiss	100—200	5—20	30—70	2—11	23—29
rocks	Marble	100—200	5—20	30—70	2—12	25—35
	Quartzite	200—400	25—30	50—90	5—15	48
Igneous	Basalt	100—300	10—15	40—80	9—14	31—38
rocks	Gabbro	100—300	10—15	40—100	6—15	—
	Granite	100—200	5—20	30—70	5—10	29—35

* Sources of references; Lama and Vutukuri (1978), Broch (1983), d'Andrea et al. (1965), Barton (1973).

In nature, intact rock masses with the mechanical properties as listed in Table 1 are of rare occurrence. Instead, rock mass conditions vary considerably, and conditions are seldom the same from site to site. Even in the same tunnel, the rock mass can vary from intact rock to swelling rock conditions. However, the rock mass can sometimes be uniform and well defined with respect to the geological structures. Examples of some structural patterns commonly encountered in underground tunnels are given below.

The design of rock reinforcement systems for a rock mass with the structural patterns shown in Figs. 11 A-D will be discussed in the next chapter.

Where rock mass conditions cannot easily be assigned to any of the examples shown in Figs. 11 A-D, additional classification systems exist. These can be used to aid the design of appropriate support requirements.

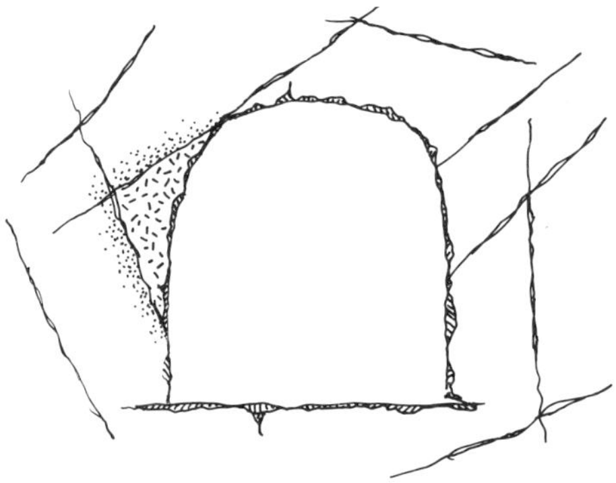

Fig. 11A
A potentially unstable wedge or block.

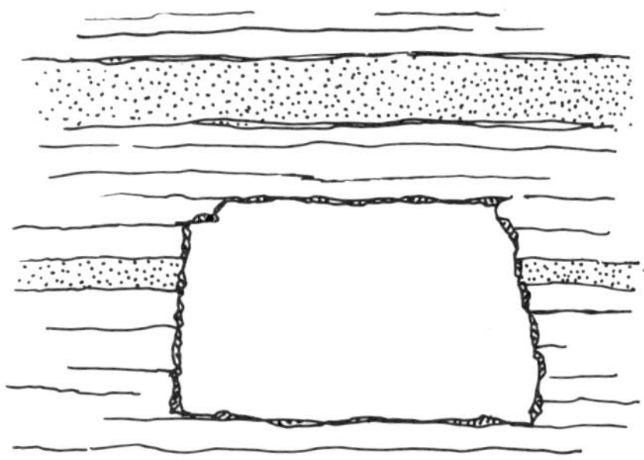

Fig. 11B
Horizontally-bedded rock overlayed by intact rock and loaded by gravity.

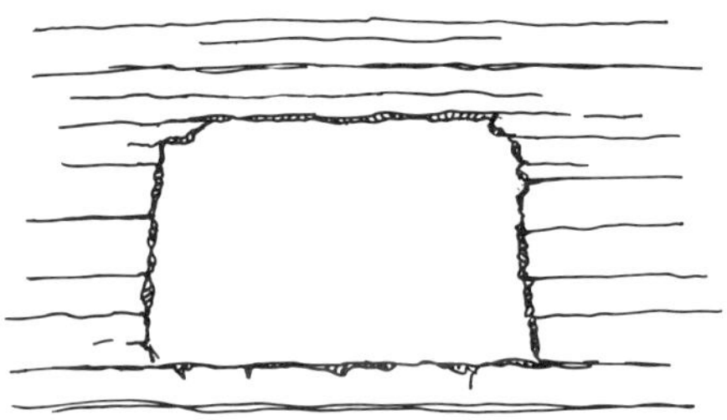

Fig. 11C
Horizontally-bedded weak rock.

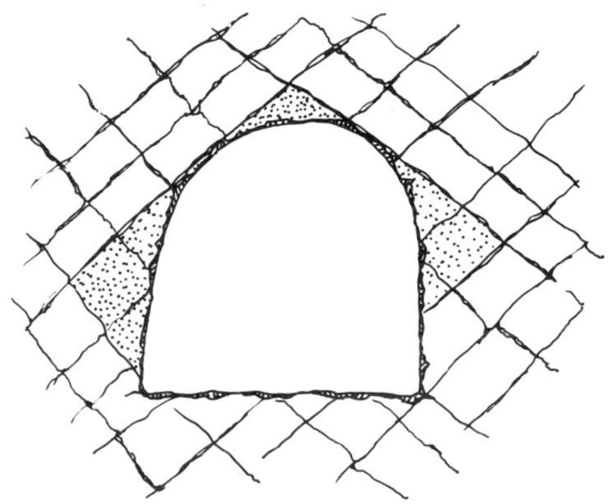

Fig. 11D
Individual blocks of rock which may loosen or fall as a result of gravity and/or in-situ stress induced loading.

6.2 Rock Mass Classification

The Geomechanics Classification proposed by Bieniawski (1973) and the *Q*-system proposed by Barton, Lien and Lunde (1974) are the two most commonly used rock mass classification systems. Both systems include sufficient information to provide a realistic assessment of factors which influence the stability of an underground excavation.

The Geomechanics Classification

The Geomechanics Classification operates with five basic classification parameters for the rock mass. The way in which these parameters have been incorporated into the Geomechanics Classification for jointed rock masses is shown in Part A of Table 2. A number of points or a rating is allocated to each range of values for each parameter and the sum of the ratings for each parameter is called the *RMR* (*R*ock *M*ass *R*ating)-value. The *RMR*-value must be adjusted for joint orientation by applying the corrections given in Part B of Table 2. An explanation of the descriptive terms used for this purpose is given in Table 3. Part C of Table 2 shows the class and description assigned to rock masses with various total ratings. The interpretation of these ratings in terms of stand-up times for underground excavations and rock mass strength parameters is given in Part D of Table 2.

The *RQD* (*R*ock *Q*uality *D*esignation) — value, DEERE (1964), referenced in Table 2, is defined as the percentage of core recovered in intact pieces of 100 mm or more in length in the total length of a borehole;

$$RQD\,(\%) = 100 \times \frac{\text{Length of core in pieces} \geqslant 100\,\text{mm}}{\text{Length of borehole}} \tag{2}$$

An *RQD*-value would usually be established for each core run of 1 or 2 metres.

Table 2
Geomechanics classification of Jointed Rock Masses. After BIENIAWSKI (1979)

A. Classification Parameters and their Ratings

<table>
<tr><th colspan="3">Parameter</th><th colspan="7">Ranges of Values</th></tr>
<tr><td rowspan="3">1</td><td rowspan="2">Strength of intact rock material</td><td>Point-load strength index</td><td>>10 MPa</td><td>4—10 MPa</td><td>2—4 MPa</td><td>1—2 MPa</td><td colspan="3">For this low range — uniaxial compressive test is preferred</td></tr>
<tr><td>Uniaxial compressive strength</td><td>>250 MPa</td><td>100—250 MPa</td><td>50—100 MPa</td><td>25—50 MPa</td><td>5—25 MPa</td><td>1—5 MPa</td><td><1 MPa</td></tr>
<tr><td colspan="2">Rating</td><td>15</td><td>12</td><td>7</td><td>4</td><td>2</td><td>1</td><td>0</td></tr>
<tr><td rowspan="2">2</td><td colspan="2">Drill core quality RQD</td><td>90 %—100 %</td><td>75 %—90 %</td><td>50 %—75 %</td><td>25 %—50 %</td><td colspan="3"><25 %</td></tr>
<tr><td colspan="2">Rating</td><td>20</td><td>17</td><td>13</td><td>8</td><td colspan="3">3</td></tr>
<tr><td rowspan="2">3</td><td colspan="2">Spacing of discontinuities</td><td>>2 m</td><td>0.6—2 m</td><td>200—600 mm</td><td>60—200 mm</td><td colspan="3"><60 mm</td></tr>
<tr><td colspan="2">Rating</td><td>20</td><td>15</td><td>10</td><td>8</td><td colspan="3">5</td></tr>
<tr><td rowspan="2">4</td><td colspan="2">Condition of discontinuities</td><td>Very rough surface
Not continuous
No separation
Unweathered wall rock</td><td>Slightly rough surfaces
Separation <1 mm
Slightly weathered walls</td><td>Slightly rough surfaces
Separation <1 mm
Highly weathered walls</td><td>Slickensided surfaces
OR
Gouge <5 mm thick
OR
Separation 1—5 mm
Continuous</td><td colspan="3">Soft gouge >5 mm thick
OR
Separation >5 mm
Continuous</td></tr>
<tr><td colspan="2">Rating</td><td>30</td><td>25</td><td>20</td><td>10</td><td colspan="3">0</td></tr>
<tr><td rowspan="4">5</td><td rowspan="3">Ground water</td><td>Inflow per 10 m tunnel length</td><td>None</td><td><10 litres / min</td><td>10—25 litres / min</td><td>25—125 litres / min</td><td colspan="3">>125 litres/min</td></tr>
<tr><td>Ratio $\frac{\text{joint water pressure}}{\text{major principal stress}}$</td><td>OR
0</td><td>OR
0,0—0,1</td><td>OR
0,1—0,2</td><td>OR
0,2—0,5</td><td colspan="3">OR
>0,5</td></tr>
<tr><td>General conditions</td><td>OR
Completely dry</td><td>OR
Damp</td><td>OR
Wet</td><td>OR
Dripping</td><td colspan="3">OR
Flowing</td></tr>
<tr><td colspan="2">Rating</td><td>15</td><td>10</td><td>7</td><td>4</td><td colspan="3">0</td></tr>
</table>

Table 2 continued

B. Rating Adjustment for Joint Orientations

Strike and dip orientations of joints		Very favourable	Favourable	Fair	Unfavourable	Very unfavourable
Ratings	Tunnels	0	—2	—5	—10	—12
	Foundations	0	—2	—7	—15	—25
	Slopes	0	—5	—25	—50	—60

C. Rock Mass Classes Determined from Total Ratings

Rating	100←81	80←61	60←41	40←21	<20
Class No.	I	II	III	IV	V
Description	Very good rock	Good rock	Fair rock	Poor rock	Very poor rock

D. Meaning of Rock Mass Classes

Class No.	I	II	III	IV	V
Average stand-up time	10 years for 15 m span	6 months for 8 m span	1 week for 5 m span	10 hours for 2,5 m span	30 minutes for 1 m span
Cohesion of the rock mass	>400 kPa	300—400 kPa	200—300 kPa	100—200 kPa	<100 kPa
Friction angle of the rock mass	>45 °	35 °—45 °	25 °—35 °	15 °—25 °	<15 °

Table 3
The Effect of Discontinuity Strike and Dip Orientations in Tunnelling. After Bieniawski (1979)

Strike perpendicular to tunnel axis				Strike parallel to tunnel axis		Dip 0 °—20 ° irrespective of strike
Drive with dip		Drive against dip				
Dip 45 °—90 °	Dip 20 °—45 °	Dip 45 °—90 °	Dip 20 °—45 °	Dip 45 °—90 °	Dip 20 °—45 °	
Very favourable	Favourable	Fair	Unfavourable	Very unfavourable	Fair	Unfavourable

Bieniawski has related his *RMR*-value to the stand-up time of an active, unsupported span. This relationship is shown in Fig. 12.

Calculation of the *RMR*-value in the field is simple when pre-investigation data is used. If a parameter value is not available, it is possible to estimate the value and to run a sensitivity check on the *RMR*-value for different values chosen.

A practical example taken from Hoek and Brown (1980), which has been modified according to the revised Geomechanics Classification, Bieniawski (1979), illustrates how the system is used:

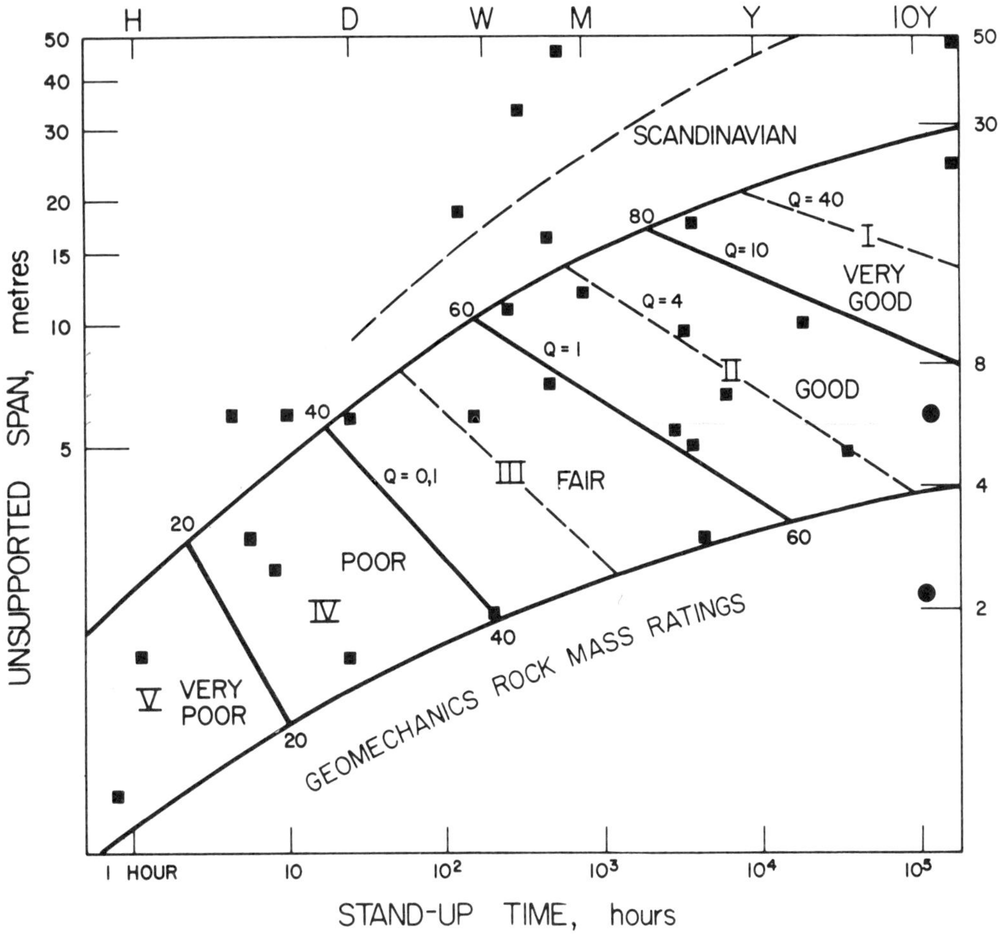

Fig. 12 Relationship between the *RMR*-value and stand-up time of an unsupported underground excavation span. After BIENIAWSKI (1979).

After the structural regions have been identified, the classification parameters for each structural region are determined from field measurements and recorded on the input data sheet given in the Appendix. Consider the example of a granitic rock mass in which a tunnel is to be driven. The classification has resulted in the following assessment;

Classification Parameter	Value or Description	Rating
1. Strength of intact material	150 MPa	12
2. *RQD*	70 %	13
3. Joint spacing	0.5 m	10
4. Condition of joint	Slightly rough surfaces. Separation < 1 mm. Slightly weathered walls.	25
5. Ground water	Dripping	4
	Total score	64

The tunnel was oriented such that the dominant joint set strikes perpendicular to the tunnel axis with a dip of 30° against the drive direction. From Table 3, this situation is described as unfavourable for which a rating adjustment of —10 is obtained from Table 2B. Thus the final rock mass rating becomes 54, which places the rock mass at the upper end of Class III with a description of fair.

Fig. 12 gives the stand-up time of an unsupported 3 metre tunnel in this rock mass as approximately 1 month.

The *Q*-System for Rock Mass Classification

On the basis of evaluation of a large number of case studies of underground excavation stability, Barton, Lien and Lunde of the Norwegian Geotechnical Institute (NGI) proposed an index for the determination of the tunnelling quality of a rock mass. The value of this index, Q is defined as follows:

$$Q = \frac{RQD \times J_r \times J_w}{J_n \times J_a \times SRF} \qquad (3)$$

where

$$RQD\,(\%) = 100 \times \frac{\text{Length of core in pieces} \geqslant 100 \text{ mm}}{\text{Length of borehole}}$$

J_n is the joint set number
J_r is the joint roughness number
J_a is the joint alteration number
J_w is the joint water reduction factor, and
SRF is a stress reduction factor

The numerical values of these parameters are determined from Table 4.

The reader may get the impression that the *Q*-system is complex and difficult to use in the analysis of practical problems. This is not the case, however. Determination of the *Q*-value for a typical rock mass will demonstrate this. It is important, however, that the comments at the end of Table 4 (these were made by Barton, Lien and Lunde) in addition to the notes listed in the tables are carefully considered.

Table 4
Classification of Individual Parameters Used in the *Q*-System. After BARTON, LIEN and LUNDE (1977).

Description	*Value*	*Notes*
1. Rock Quality Designation	*(RQD)*	
A. Very poor	0—25	(1) Where *RQD* is reported or measured as ⩽10, (including 0) a nominal value of 10 is used to evaluate *Q* in equation (3).
B. Poor	25—50	
C. Fair	50—75	
D. Good	75—90	
E. Excellent	90—100	(II) *RQD* intervals of 5, i.e. 100,95,90, etc. are sufficiently accurate.
2. Joint Set Number	(J_n)	
A. Massive, no or few joints	0.5—1.0	
B. One joint set	2	
C. One joint set plus random	3	
D. Two joint sets	4	
E. Two joint sets plus random	6	
F. Three joint sets	9	(I) For intersections use ($3.0 \times J_n$)
G. Three joint sets plus random	12	(II) For portals use ($2.0 \times J_n$)
H. Four or more joint sets, random, heavily jointed, "sugar cube" etc.	15	
J. Crushed, rock, earthlike	20	
3. Joint Roughness Number	(J_r)	
(a) Rock wall contact and		
(b) Rock wall contact before 10 cms shear		
A. Discontinuous joints	4	
B. Rough or irregular, undulating	3	
C. Smooth, undulating	2	
D. Slickensided, undulating	1.5	(I) Descriptions refer to small scale features and intermediate scale features, in that order.
E. Rough or irregular, planar	1.5	
F. Smooth, planar	1.0	(II) Add 1.0 if the mean spacing of the relevant joint set is greater than 3 m.
G. Slickensided, planar	0.5	
c) No rock wall contact when sheared		(III) J_r = 0.5 can be used for planar slickensided joints having lineations, provided the lineations are orientated for minium strength
H. Zone containing clay minerals thick enough to prevent rock wall contact	1.0	
J. Sandy, gravelly or crushed zone thick enough to prevent rock wall contact	1.0	
4. Joint Alteration Number	(J_a)	**(ϕ_r approx.)**
(a) Rock wall contact		
A. Tightly healed, hard non-softening, impermeable filling i.e. quartz or epidote	0.75	(—)
B. Unaltered joint wall, surface staining only	1.0	(25—35°)
C. Slightly altered joint walls. Non-softening mineral coatings, sandy particles, clay-free disintegrated rock etc.	2.0	(25—30°)
D. Silty-, or sandy-clay coatings, small clay fraction (non-soft.)	3.0	(20—25°)

	Description	Value		Notes
		(J_a)	**(ϕ_r approx.)**	
E.	Softening or low friction clay mineral coatings, i.e. kaolinite or mica. Also chlorite, talc, gypsum, graphite etc., and small quantities of swelling clays.	4.0	(8—16°)	
	(b) Rock wall contact before 10 cms shear			
F.	Sandy particles, clay-free disintegrated rock etc.	4.0	(25—30°)	
G.	Strongly over-consolidated non-softening clay mineral fillings (continuous, but <5 mm thickness).	6.0	(16—24°)	
H.	Medium or low over-consolidation, softening, clay mineral fillings. (continuous but <5 mm thickness).	8.0	(12—16°)	
J.	Swelling clay fillings, i.e. montmorillonite (continuous, but <5 mm thickness) Value of J_a depends on percent of swelling clay-size particles, and access to water etc.	8—12	(6—12°)	
	(c) No rock wall contact when sheared			
K, L, M	Zones or bands of disintegrated or crushed rock and clay (see G,H,J for description of clay condition).	6,8, or 8—12	(6—24°)	
N.	Zones or bands of silty-or sandy-clay, small clay fraction (non-softening).	5.0	(—)	
O,P, R.	Thick, continuous zones or bands of clay (see G,H,J, for description of clay condition).	10,13, or 13—20	(6—24°)	
5.	**Joint Water Reduction Factor**	(J_w)	**Approx. water pres. (kp/cm²)**	
A.	Dry excavations or minor inflow, i.e. <5 1/min. locally.	1.0	<1.0	
B.	Medium inflow or pressure, occasional outwash of joint fillings.	0.66	1—2.5	
C.	Large inflow or high pressure in competent rock with unfilled joints	0.5	2.5—10	(I) Factors C to F are crude estimates. Increase J_w if drainage measures are installed.
D.	Large inflow or high pressure, considerable outwash of joint fillings	0.33	2.5—10	
E.	Exceptionally high inflow or water pressure at blasting, decaying with time	0.2—0.1	>10	
F.	Exceptionally high inflow or water pressure continuing without noticeable decay	0.1—0.05	>10	(II) Special problems caused by ice formation are not considered.
6.	**Stress Reduction Factor**			
	(a) Weakness zones intersecting excavation, which may cause loosening of rock mass when tunnel is excavated		*(SRF)*	
A.	Multiple occurrences of weakness zones containing clay or chemically disintegrated rock, very loose surrounding rock (any depth).		10	(I) Reduce these values of *SRF* by 25—50% if the relevant shear zones only influence but do not intersect the excavation.
B.	Single weakness zones containing clay or chemically disintegrated rock (depth of excavation <50 m).		5	
C.	Single weakness zones containing clay or chemically disintegrated rock (depth of excavation >50 m).		2.5	
D.	Multiple shear zones in competent rock (clay-free), loose surrounding rock (any depth),		7.5	
E.	Single shear zones in competent rock (clay-free) (depth of excavation <50 m).		5.0	
F.	Single shear zones in competent rock (clay-free) (depth of excavation >50 m).		2.5	
G.	Loose open joints, heavily jointed or "sugar cube" etc. (any depth).		5.0	

	Description	*Value*			*Notes*
	(b) Competent rock, rock stress problems				
		σ_c/σ_1	σ_t/σ_1	*(SRF)*	
H.	Low stress, near surface	>200	>13	2.5	(II) For strongly anisotropic virgin stress field (if measured): when $5 \leqslant \sigma_1/\sigma_3 \leqslant 10$, reduce σ_c and σ_t to 0.8 σ_c and 0.8 σ_t. When $\sigma_1/\sigma_3 > 10$, reduce σ_c and σ_t to 0.6 σ_c and 0.6 σ_t, where: σ_c = unconfined compression strength, and σ_t = tensile strength (point load), and σ_1 and σ_3 are the major and minor principal stresses.
J.	Medium stress	200—10	13—0.66	1.0	
K.	High stress, very tight structure (usually favourable to stability, may be unfavourable for wall stability)	10—5	0.66—.33	0.5—2	
L.	Mild rock burst (massive rock)	5—2.5	0.33—.16	5—10	
M.	Heavy rock burst (massive rock)	2.5	0.16	10—20	(III) Few case records available where depth of crown below surface is less than span width. Suggest SRF increase from 2.5 to 5 for such cases (see H).
	(c) Squeezing rock: plastic flow of incompetent rock under the influence of high rock pressure				
N.	Mild squeezing rock pressure			5—10	
O.	Heavy squeezing rock pressure			10—20	
	(d) Swelling rock: chemical swelling activity depending one presence of water				
P.	Mild swelling rock pressure			5—10	
R.	Heavy swelling rock pressure			10—15	

Additional Notes on the Use of the Above Table

When making estimates of the rock mass quality (Q) the following guidelines should be followed, in addition to the notes listed in the tables:

1. When borehole core is unavailable, RQD, can be estimated from the number of joints per unit volume, in which the number of joints per metre for each joint set are added. A simple relation can be used to convert this number to RQD for the case of clay free rock masses:

$$RQD = 115 - 3.3\,J_v \text{ (approx.) where } J_v = \text{total number of joints per m}^3$$

$$(RQD = 100 \text{ for } J_v < 4.5)$$

2. The parameter J_n representing the number of joint sets will often be affected by foliation, schistosity, slaty cleavage or bedding etc. If strongly developed these parallel "joints" should obviously be counted as a complete joint set. However, if there are few "joints" visible, or only occasional breaks in the core due to these features, then it will be more appropriate to count them as "random joints" when evaluating J_n.
3. The parameters J_r and J_a (representing shear strength) should be relevant to the *weakest significant joint set or clay filled discontinuity* in the given zone. However, if the joint set or discontinuity with the minimum value of (J_r/J_a) is favourably oriented for stability, then a second, less favourably oriented joint set or discontinuity may sometimes be more significant, and its higher value of J_r/J_a should be used when evaluating Q. *The value of J_r/J_a should in fact relate to the surface most likely to allow failure to initiate.*
4. When a rock mass contains clay, the factor *SRF* appropriate to *loosening loads* should be evaluated. In such cases the strength of the intact rock is of little interest. However, when jointing is minimal and clay is completely absent the strength of the intact rock may become the weakest link, and the stability will then depend on the ratio rock-stress/rock-strength. A strongly anisotropic stress field is unfavourable for stability and is roughly accounted for as in note II in the table for stress reduction factor evaluation.
5. The compressive and tensile strengths (σ_c and σ_t) of the intact rock should be evaluated in the saturated condition if this is appropriate to present or future in situ conditions. A very conservative estimate of strength should be made for those rocks that deteriorate when exposed to moist or saturated conditions.

To relate the Q-value to the behaviour and support requirements of an underground excavation, BARTON, LIEN and LUNDE have defined a quantity of the excavation called the **equivalent dimension** D_e. This dimension is obtained by dividing the span, diameter or wall height of the excavation by a quantity called the **excavation support ratio,** *ESR*. The quantity D_e is defined as follows:

$$D_e = \frac{\text{Excavation span, diameter or height (m)}}{\text{Excavation Support Ratio } (ESR)} \tag{4}$$

The numerical value of the excavation support ratio is related to the intended use of the excavation and to the degree of safety demanded. BARTON (1976) has given the following values for *ESR*:

	Excavation category	*ESR*
A.	Temporary mine openings	3—5
B.	Permanent mine openings, water tunnels for hydropower (excluding high pressure penstocks) pilot tunnels, drifts and headings for large excavations	1.6
C.	Storage rooms, water treatment plants, minor road and railway tunnels, surge chambers, access tunnels	1.3
D.	Power stations, major road and railway tunnels, civil defence chambers, portals, intersections	1.0
E.	Underground nuclear power stations, railway stations, sports and public facilities, factories	0.8

The relationship between the Q-value and the equivalent dimension D_e of an excavation which will stand unsupported is illustrated in Fig. 13.

The use of Fig. 13 will be reviewed in more detail in the next chapter, when reinforcement design is discussed.

A practical example taken from HOEK and BROWN (1980) illustrates how the Q-system for rock mass classification is used:

> An underground crusher station is to be excavated in the limestone footwall of a lead-zinc orebody. It is required to determine the span which can be left unsupported.

The analysis is performed as follows:

	Item	Description	Value
1.	Rock Quality	Good	$RQD = 80\%$
2.	Joint sets	Two sets	$J_n = 4$
3.	Joint roughness	Rough	$J_r = 3$
4.	Joint alteration	Clay gouge	$J_a = 4$
5.	Joint water	Large inflow	$J_w = 0.33$
6.	Stress reduction	Medium stress	$SRF = 1.0$

These values, substituted in eq. (3) give

$$Q = \frac{80}{4} \times \frac{3}{4} \times \frac{0.33}{1} = 5$$

From Fig. 13, the maximum equivalent dimension D_e for an unsupported excavation in this rock mass is 4 metres. A permanent underground mine opening has an excavation support ratio *ESR* of 1.6 and hence, the maximum unsupported span which can be considered for the crusher station is $ESR \times D_e = 1.6 \times 4 = 6.4$ metres.

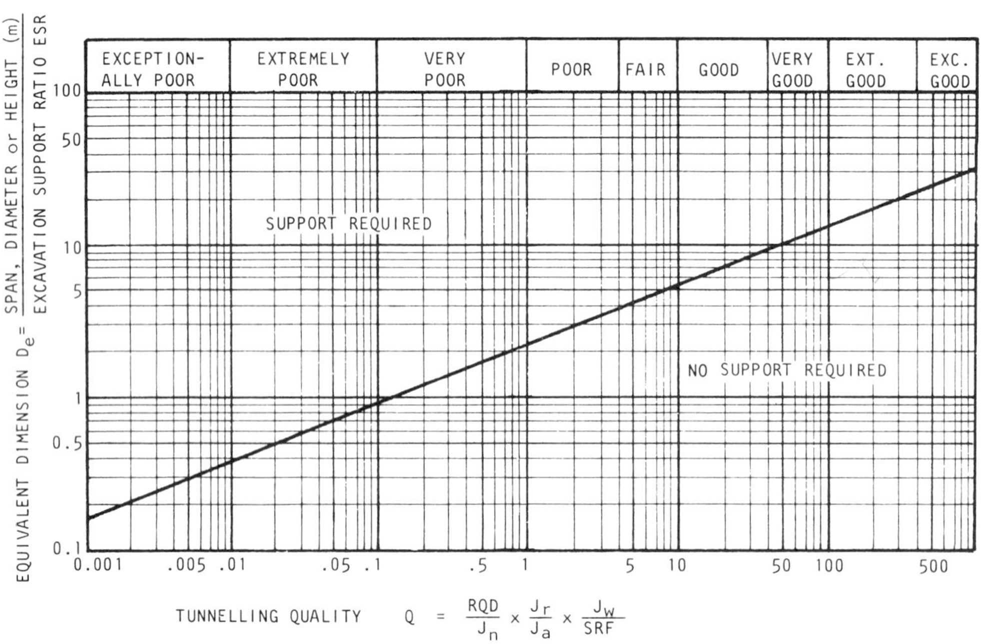

Fig. 13 Relationship between the maximum equivalent dimension (D_e) of an unsupported underground excavation and the rock mass quality (Q). After BARTON, LIEN and LUNDE (1977).

It is common practice that both systems, the Geomechanics Classification and the *Q*-system, are used in the solution of a problem. It has been found that with problems involving extremely weak ground which results in squeezing, swelling or flow conditions, the Geomechanics Classification is difficult to apply. Here, the *Q*-system is therefore recommended.

6.3 Special Conditions

High in-situ stresses

The in-situ rock stress conditions can be determined by rock stress measurements. The magnitudes, directions and the relation between magnitudes of the principal stresses, together with the mechanical properties of the rock mass guide the engineer in his prediction of the likelihood of the occurrence of rock burst or spalling problems associated with excavation. The behaviour of the rock mass around an underground excavation is, to a great extent, related to the in-situ stresses. When an underground excavation is developed, the virgin, in-situ stresses are affected to distances of approximately one "excavation diameter" from its periphery. The vertical stresses in the rock mass are, in general, a function of the height of the overburden times the rock mass density. The horizontal stresses vary considerably. They can, if tectonically induced, reach a magnitude of the same order as the strength of the rock mass. In highly stressed areas in the walls of an excavation, large tangential stress concentrations may arise, which exceed the strength of the rock mass. Spalling or rock burst will then occur.

In Fig. 14 some geological conditions that can result in rock bursts are illustrated.

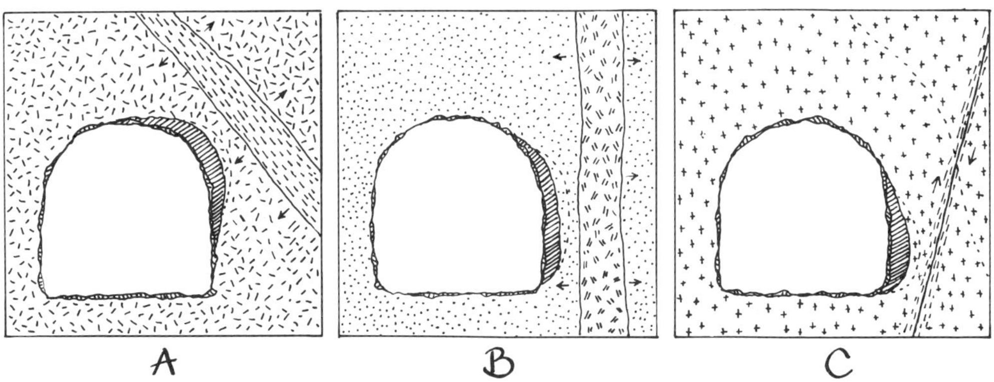

Fig. 14 Examples of geological conditions that can result in rock burst. Dashed areas indicate rock which is subjected to potential rock burst. After WAHLSTRÖM (1973).
A: Load of overlying rock in a deeply located tunnel in brittle granite causes semiplastic movement in a weak lens of mica-rich schist. Rock burst (hatched) occurs in the tunnel at a point closest to the lens.
B: A dike of igneous rock which is forcefully injected into a hard quartzite is accompanied by residual stresses that will promote rock burst in the wall of the tunnel.
C: Forces that have caused elastic strain in strong igneous rock are directed towards the tunnel opening to cause rock burst.

Rock burst or spalling problems will also affect the choice of the rock reinforcement system. Experience from very deep mining in hard rock has shown that the immediate rock surrounding the excavation will inevitably be severely fractured. To control the movement of this fractured rock, if large in-situ stresses persist, an elastic support system is required, ORTLEPP and GAY (1984). Even where rock burst problems are less severe, and do not necessarily result in a complete fracturing of the surrounding rock, an elastic support system is recommended. Hence, provided that the fractured rock can be kept in place to prevent progressive ravelling, there is little danger of the fracture zone propagating to the point where the excavation would collapse. Untensioned rockbolts have proved to be very effective in the prevention of ravelling and spalling. The rockbolts are, if installed close to the advancing face, tensioned by the deformation of the rock mass as the support provided by the face is removed and the rock is allowed to relax.

Swelling and squeezing rock

A rock mass with joint filling material, which swells in the presence of water, is a common problem in underground excavations and no quantitative design techniques are readily available. The same problem applies to "squeezing" rock. Squeezing rock means that the rock surrounding the excavation displays time-dependent deformations. Here, the rock stresses have exceeded the strength of the rock mass or the creep limit, and the rock mass deforms plastically.

One approach to the design of the reinforcement which applies both to the problem of swelling and squeezing rocks is to allow the opening to deform without any applied support. In squeezing rock, the deformations that should be allowed before any support measures are taken can be of the order of 5—10% of the tunnel diameter. The problem with this approach is the determination of the magnitude of deformation that can be permitted before significant rock mass strength is lost. As an alternative to rock support under swelling conditions, the rock could be sealed immediately after excavation by the application of shotcrete.

In Fig. 15 some geological conditions that promote ground squeezing in tunnels are illustrated.

Swelling or squeezing rocks usually have low rock mass strength properties. Mechanically anchored rockbolts are therefore considered to be inappropriate for these rock types. It is recommended that tensioned resin or cement grouted rockbolts are used here.

A complete reinforcement system for an underground excavation in swelling or squeezing rock may include an initial layer of shotcrete, followed by tensioned rockbolts once the ground has been allowed to deform. It is possible that cracks will form in the initial layer of shotcrete. Small rock fragments could then fall out

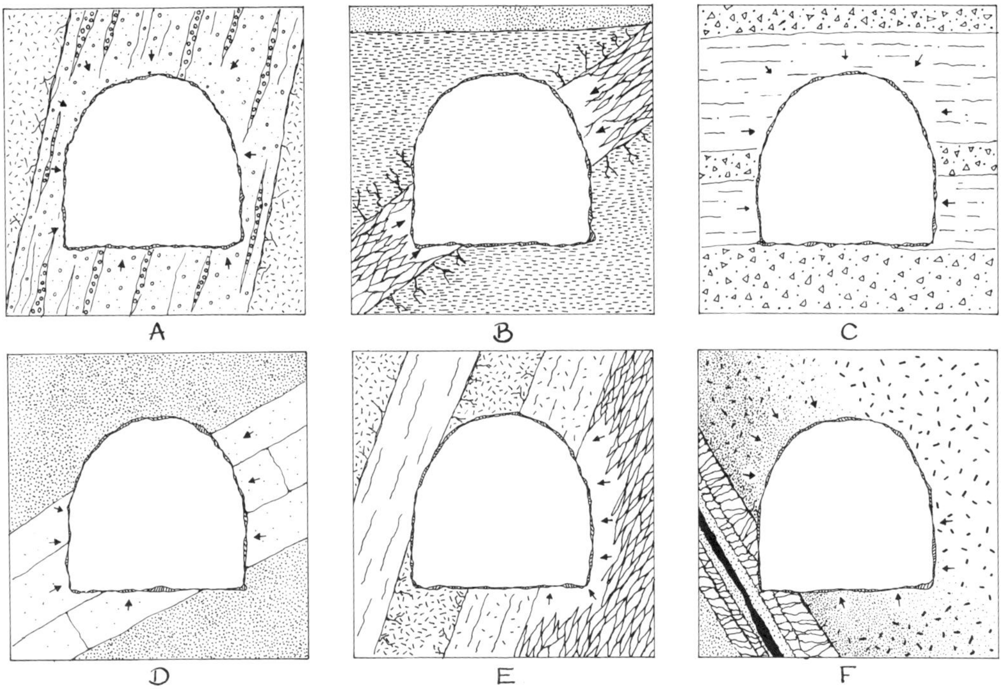

Fig. 15 Examples of geological conditions that promote ground squeezing in tunnels. Arrows indicate direction of movement of ground squeezing into the tunnel opening. After Wahlström (1973).

A: The tunnel is located in a wide shear zone in granite. The rock mass has been thoroughly crushed. It consists of a gouge matrix containing lenses of angular rock fragments. The gouge is damp but it is essentially impermeable to water penetration.

B: A flat fault in brittle shale has produced gouge containing a braided pattern of shears. Swelling is localized at exposures of fault.

C: Layer of altered volcanic ash (bentonite) contains abundant expanded montmorillonite. Load pressure of overlying rocks causes squeezing.

D: A soft, poorly compacted mudstone layer is squeezed into the tunnel by the weight of the overlying rocks.

E: Closely sheared, altered biotite schist contains abundant chlorite and clay minerals and squeezes into the tunnel.

F: An extensively altered fine-grained porphyry adjacent to a vein tends to squeeze into the tunnel.

between the bolts. A layer of weldmesh attached to the bolts and anchored at intermediate points to the shotcrete is therefore recommended here. A final layer of shotcrete, 25 to 50 mm thick and placed over the weldmesh gives a total layer of 50 to 75 mm of reinforced shotcrete, together with the tensioned grouted anchored rockbolts to support the excavation.

Groundwater

Groundwater can cause serious problems in underground excavations. It can be very difficult, for example, to predict the groundwater situation in a tunnel which is driven through different geological formations. The flow of groundwater into an underground excavation has particular significance for the analysis of the stability of the excavation and the design of the rock reinforcement.

A continuous flow of groundwater from boreholes that are to be used for rock bolting normally excludes the use of resin or cement grouted rockbolts. Mechanically or friction anchored rockbolts should only be used for temporary reinforcement of the rock. Drainage of near-surface zones of rock is usually provided if the site investigation indicates the presence of groundwater. However, even with the provision of drainage precautions in connection with the structural stability of the excavation, it is advisable to make some allowance for water pressures. If shotcrete is applied, it is important that this is properly drained.

Cement grouting, (injection of cement grout into the rock) is sometimes used in combination with rockbolts to reduce the rock mass permeability and increase its strength. When the grouting is properly designed and constructed, effective results can be obtained.

When the presence of groundwater is anticipated in the development of an underground excavation, it is of crucial importance that the pre-investigation scheme is directed towards the identification of the major groundwater bearing zones. Appropriate precautions must be taken in the excavation technique to meet the expected problem.

Dynamic effects

In general, "dynamic effects" refer to earthquake and blasting activity, although waterhammer effects in hydraulic tunnels may be significant in some instances. Although there are techniques available for the assessment of dynamic effects on rock bocks, they are of limited applicability, since a number of assumptions must be made for the evaluation of the results. The displacements from the dynamic loading of the rock mass surrounding an underground excavation are in general small, but sometimes the progressive movement from round-by-round blasting can cause detrimental loosening. The effect of blasting on mechanically anchored rockbolts cannot be ignored. It must be considered in the design of the reinforcement system. To ensure that the tension in a mechanically anchored rockbolt is maintained the bolt should not be installed closer than 10 m to a nearby blast, as a "rule of thumb".

Pull-out test results in hard rock, of fully cement grouted cable bolts, have shown that if the reinforced rock is subjected to the dynamic loading from a nearby production blast, the load bearing capacity of the cable will only be significantly affected, when the particle velocity value exceeds 500 mm/s close to the cable, Stillborg (1984).

Cautious blasting

An essential step in good rock reinforcement practice is to minimize overbreak and adopt a technique of cautious blasting in underground excavations. To minimize overbreak, it is essential that the direction of the blastholes is carefully selected and that the borehole deviation and the charging operation are controlled.

The importance of a controlled drilling and blasting procedure was demonstrated in the Luossavaara full scale mining trial. In the open stope operation 100 m long and 165 mm (6½″) large blastholes were used for the production hole drilling. The very stable conditions encountered in the hanging-wall of the stopes during the mining, were partly due to the pre-reinforcement of the hanging-wall with long, fully grouted cable bolts. The main reason for the stable conditions was, however, due to the use of an advanced drilling technique which enabled the boreholes to be drilled very accurately, in combination with a system of continuous sampling of the cuttings. This enabled the mining engineers to both avoid charging holes or portions of holes that were located outside the orebody, and also to make special charging arrangements for boreholes close to the hanging-wall contact, Stillborg (1985).

Cautious blasting involves drilling a number of closely spaced parallel boreholes along the final excavation surface, placing low density decoupled charges in the holes and detonating these charges together, but after the detonation of the cut and main blastholes. This is also sometimes called "smooth blasting". Figure 16 illustrates the difference in damage intensity of the rock surrounding a tunnel in the case of conventional blasting and smooth blasting. In Fig. 16, the extent of the blast damage for some common explosives is also shown.

In development workings the number of boreholes through moderately jointed quartz-bearing porphyry, for example, was increased by 35 to 40%. Here, it was possible to conduct specified profiling and to minimize blast damage to the surrounding rock. Systematic bolting had been prescribed previously. However, with the cautious blasting, only occasional spot bolting was now necessary, Stillborg (1983).

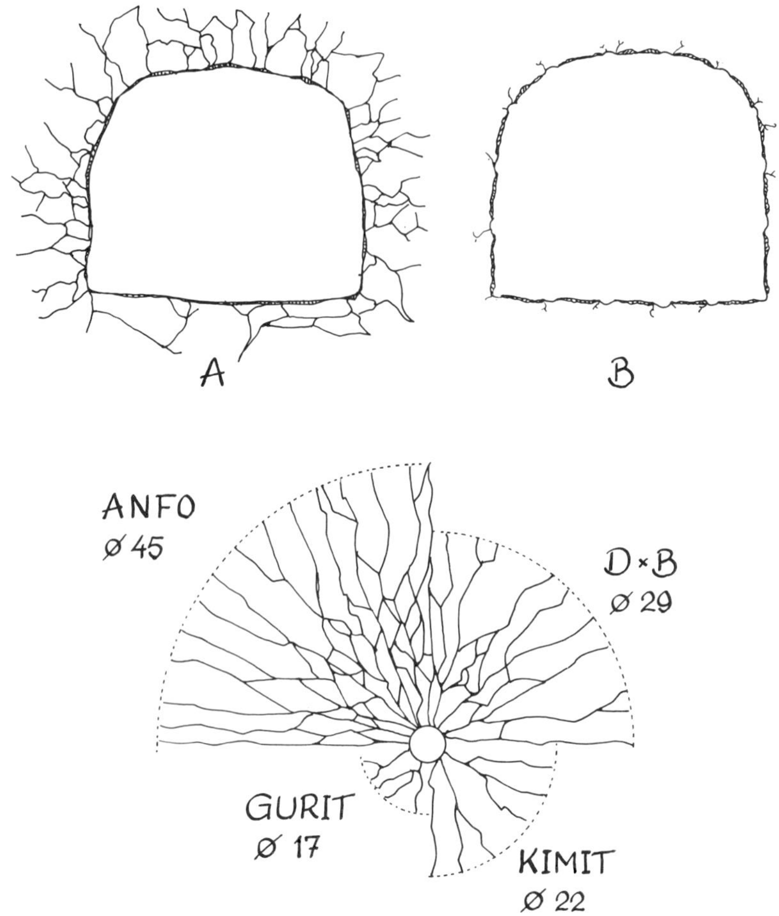

Fig. 16 Schematic illustration of the difference in damage intensity of the rock surrounding a tunnel in the case of conventional blasting (A) and smooth blasting (B). The extent of the blast damage for some common explosives is also shown. Blasthole diameter is given in mm. After Holmberg et al. (1983).

ANFO — ANFO
DxB — Plastic Dynamite 28% NG/EGDN
Kimit — Watergel (cap sensitive)
Gurit — Low strength powdered NG/EGDN sensitized explosive

6.4 Choice of Rockbolt System

It is beyond the scope of this handbook to present the type of rockbolt/cable bolt system to be used to meet every combination of conditions present in an underground excavation. However, the following guideline should prove useful, if the inappropriate rockbolt or cable bolt system for a given situation is to be avoided.

Mechanically anchored rockbolts are not recommended for use under the following conditions;
— in soft rock conditions where the rock type or the joint filling material may affect the gripping force of the anchor
— near blasting activities, where the tension in a bolt may be lost
— in permanent reinforcement systems, unless the bolt is grouted
— close to an advancing face where subsequent rock mass deformation is significant
— in areas of large rock stresses where rock burst or spalling problems may be encountered
— in applications where rockbolt tension cannot be checked
— to resist shear movements of the rock
— in very hard rock
— in very fractured rock.

Grouted rockbolts/cable bolts are not recommended for use under the following conditions;
— in applications where the quality of the grouting agent cannot be checked
— in boreholes with a continuous flow of groundwater
— when cement is the grouting agent and immediate support action is necessary
— in rocks having open joints and/or voids, unless the grouting can be checked.

Untensioned grouted rockbolts/cable bolts are not recommended for use under the following condition;
— in rock masses which have been subjected to extensive deformation.

Tensioned grouted rockbolts/cable bolts are not recommended for use under the following conditions;
— close to an advancing face where subsequent rock mass deformation is significant
— in areas of large rock stresses, where rock burst or spalling problems may be encountered.

Friction anchored rockbolts are not recommended for use under the following conditions;
— in permanent reinforcement systems unless special corrosion protection is applied
— to resist shear movements of the rock.

Split Set rockbolts are not recommended for use under the following conditions;
— in narrow or confined areas
— where borehole diameters are difficult to control
— in very fractured or soft rock conditions.

Swellex standard rockbolts are not recommended for use under the following condition;
— in areas where large deformations can be anticipated.

Chapter 7

Design of Rock Reinforcement

The mechanical behaviour of rock masses reinforced by rockbolts or cable bolts, (i.e. the "rock mass — rockbolt interaction"), is still, in general, incompletely understood. This is also true of the rockbolt/cable bolt itself as a constructional element, in spite of the widespread use of rockbolts and cables. This is reflected in the lack of appropriate design criteria for rockbolt and cable bolt reinforcement. For long cable bolts, (cables longer than 6 metres), the design of the reinforcement systems also, at present, relies heavily on the experience of the responsible engineer and/or empirical results from previous trials. Nevertheless, an attempt is made in this chapter to compile some of the more useful design concepts. These are commonly applied to practical rock reinforcement problems, where rockbolts or cable bolts are used as reinforcing elements.

7.1 Rock Mass — Rockbolt Interaction

It is important to appreciate the qualitative differences in interaction between mechanically anchored rockbolts, fully grouted bolts and friction anchored bolts and a rock mass.

— In mechanically anchored bolts, the strain is constant along the free length of the bolt. Any rock displacement between the two anchor points of the bolt, the face plate and the expansion shell, will be distributed over the complete length of the bolt shank. This makes the expansion shell anchored rockbolt a flexible rock reinforcement system.

— In fully grouted bolts the load is transferred from the grout to the rock, and the deformation of the rock mass and the reinforcement cannot be separated. The load is distributed over a limited distance from the rock joint (approximately 5—20 bolt diameters), which makes the fully grouted bolt a stiff rock reinforcement system. This applies to both the rockbolt and the cable bolt.

— Friction anchored rockbolts are similar to the fully grouted bolts in that the deformation of the rock mass and the bolt cannot be separated. This is only true for the Swellex rockbolt. The Split Set will slip before the ultimate load bearing capacity of the steel tube is reached. The Swellex will not slip. It will continue to behave like a fully grouted rockbolt until it finally breaks.

A commonly discussed problem in the design of rockbolt/cable bolt systems is whether tensioned bolts should be used in preference to untensioned bolts. No simple answer to this question exists. However, some general guidelines can be followed. Tensioned bolts together with the rock behave, in principle, like a pre-stressed concrete element. This behaviour may be compared with untensioned bolts, which together with the rock, act like a fractured reinforced concrete structure.

Whenever tensioned rockbolts are used in a randomly jointed rock mass the individual blocks of rock interact to create a reinforced rock arch. Excessive deformations do not occur. However, untensioned fully grouted rebars and cables installed close to an advancing face will be tensioned by the subsequent relaxation of the rock mass and tensioned bolts are not required.

In most rockbolting, except in the case of support of unstable blocks of rock when the bolts or cables are designed to hold the dead weight of the block, the load taken by the bolts is small compared with the loads acting in the rock. The rockbolts enable the rock mass to be "self supporting".

To assess the relevance of rock bolting, it is necessary to consider the rockbolt-rock mass interaction. The load-deformation characteristics of the rock mass must be determined with some degree of confidence before a ground support system can be considered. When the rock mass displays well defined structural patterns like the examples given in section 6.1, rockbolts or cables are recommended for reinforcement of the rock. Rockbolt or cable reinforcement systems are especially recommended in situations where unstable wedges or rock blocks, free to fall or slide under their own weight are present. Their design is simpler than that of steel sets, for example, whenever eccentric loading is present.

In a more complex, stressed structure of rock, the stiffness and the timing of installation of reinforcement becomes increasingly important. It should be compatible with the response of the rock mass on unloading, the so called "ground reaction curve". If rockbolts or cable bolts are installed before relaxation of the rock mass has occurred, they may attract excessive loading, if the type of bolts selected cannot accommodate the associated deformation. However, if the bolts are installed after most of the relaxation has occurred, and the bolts are not sufficiently stiff, plastic deformation of the rock may continue and as a result be too large for the bolts to be effective. The concept is illustrated in Fig. 17.

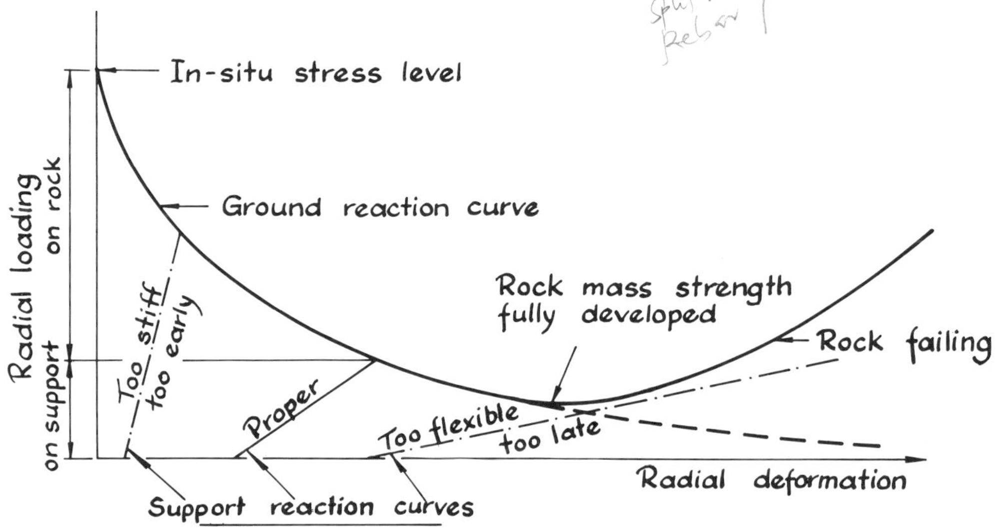

Fig. 17 Rock mass — rock reinforcement interaction concept. Partly after RABCEWICZ and GOLSER (1973).

7.2 Design Principles

The rockbolt/cable bolt reinforcement system should be designed to minimize deformations induced by the dead weight of loosened rock, as well as those induced by stress redistribution in the rock surrounding an excavation.

The basic approach to the design of rock reinforcement is to stabilize surface blocks and/or rock formations by restricting relative displacements of adjacent blocks of rock. For large openings which require long cables in particular, this will create confinement to the back.

Apart from their construction and load bearing capacity, there is no basic difference between the grouted rockbolt and the grouted cable. The general design principles for rockbolts and cable bolts should not therefore be separated. The design of rock reinforcement using grouted rockbolts considered in this chapter, is equally applicable to the design of rock reinforcement using grouted cable bolts.

In general, the stability of a near surface underground excavation, low stress environment, depends on the structural conditions in the rock mass. However, at depth, the relationship between the stress in the rock and the strength of the rock determines the stability of the excavation. For deep level mining, where rock stresses locally exceed the inherent strength of the rock, design methods for rockbolting are less developed than those used for near surface excavations in a low stress environment. For low stresses (relative to the rock mass strength), the structures in the rock effectively determine the stability of the excavation. The lower the rock stresses are, the more important the structures become for the stability of the excavation. In a low stress environment, no rock burst or spalling will occur. On the other hand, low rock stresses cannot generate effectively, either an increase in the strength of the structures or confinement to the rock mass. This implies that the size and orientation of structures intersecting the opening become important in the estimation of the size of the plastic zone and hence the analysis of structural stability for rockbolt reinforcement.

When the rockbolts are designed for temporary reinforcement of the rock, deformation measurements become essential for the selection and timing of suitable permanent reinforcement measures. This philosophy is adopted by e.g. RABCEWICZ and GOLSER (1973) in the NATM, (the New Austrian Tunnelling Method).

The design of rockbolt systems has been divided into the following:

— Analysis of structural stability
— Empirical assessments
— Numerical models.

Whatever design method is selected, the following points should be considered:

— Design work should continue throughout the construction period, making use of monitoring results.
— A good balance between economy and safety should be maintained in the selection of the rock reinforcement method which will be applied.
— Theoretical procedures should be tempered by experience and engineering judgement.

7.2.1 Analysis of Structural Stability

Although the stability of a near surface underground excavation is in general controlled structurally, especially in hard rock conditions, stress analysis may still play an important part in the assessment of the stability of the opening. Zones around the excavation, subjected to potential destressing or tension slip along low-strength joint surfaces, and stress-induced instability of rock blocks are of major design importance and must be considered.

Considering the four examples of jointed rock masses given in section 6.1, for each type of rock mass, the structural stability of an excavation may be assessed by using one of the following three approaches:

1. Limited rock block stability analysis.
2. The beam or slab concept for bedded rock.
3. The rock arch concept.

Limited rock block stability analysis

Limited rock block stability analysis is applicable to unstable wedges or blocks of rock which are free to fall or slide under their own weight. This will probably occur independently of the remaining rock mass.

The determination of the volume of a loose wedge or block of rock presents a difficulty in the application of the analysis in practice. The volume must be estimated from location, size and orientation of structures that define the outline of the block.

For the determination of the shape and volume/weight, as well as the sliding direction, when applicable, of rock wedges or blocks in the roof or walls of a tunnel, the stereographic projection technique described by e.g. HOEK and BROWN (1980) is recommended. Once the appropriate information has been obtained, using a stereographic projection technique, the number of rockbolts required to hold the block can be calculated. One of the two methods described below can be used. In the first case, the support of rock wedges or blocks, free to fall are considered. In the second case the support of rock wedges or blocks, free to slide are considered.

Support of wedges or blocks of rock which are free to fall:

Here, the frictional resistance of the joints is not considered in the design of the support system, since only the dead weight of the loose rock is to be supported, see Fig. 18.

The determination of the length of the rockbolts should be based on the dimensions of the loose block. Due consideration should be given to the degree of certainty within which the block size has been determined. The bolt length should be long enough to ensure that, within certain limits, the portions of the bolts anchored inside the solid rock will be capable of holding the weight of the block they are designed to support. This applies to the fully grouted and friction anchored rockbolts. A fully cement grouted rebar should extend approximately 1 m into solid rock. It is, however, important to check that sufficient anchoring length for the bolt type considered has been allocated, using pull-out tests, since local conditions may affect the necessary anchoring length, see also section 5.1.

The author is aware of a number of cases where rock blocks have failed, even though a safety factor of two or more was apparently used in the design of the reinforcement system. The explanation of these failures may be as follows. The block, as it fails, does not load the bolt system uniformly. Some of the bolts are overstressed, and break. The remaining bolts cannot support the weight of the block. The block falls. In some cases this explanation is supported by the fact that the block has fallen in a "cantilever action", rotating through 90° before hitting the floor.

These failures emphasize the importance of correct rock block shape and weight definition, as well as an appropriately defined potential mode of failure.

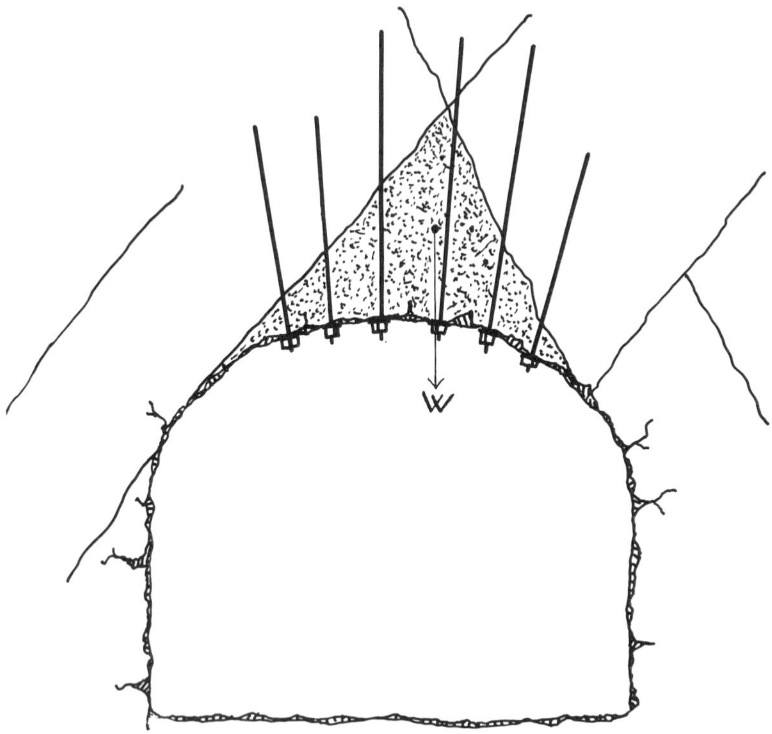

We have

$$N = \frac{W \times f}{B} \qquad (5)$$

where

N = number of rockbolts
W = weight of wedge, (density of rock approximately 2.7 t/m^3)
f = safety factor, usually $2 \leqslant f < 5$
B = load bearing capacity of the bolt, see also section 5.1

Fig. 18 Reinforcement of a wedge of rock free to fall under its own weight.

Support of wedges or blocks of rock which are free to slide:

In this case, the frictional resistance of the sliding surfaces must be considered in the design of the reinforcement system. Most rock mass structures, in hard as well as in soft rock conditions have rough and uneven joint surfaces. If the joint surfaces are properly matched, the joint can sustain large shear forces. The frictional resistance to shear will increase with the load normal to the joint surface. However, if the joint surfaces are not properly matched, relatively large shear displacements, of the order of several mm must occur before maximum frictional resistance to sliding is attained. If tensioned bolts are placed roughly perpendicular to the joint surface, the resulting normal load on the joint surface will effectively increase the shear strength of the rock mass.

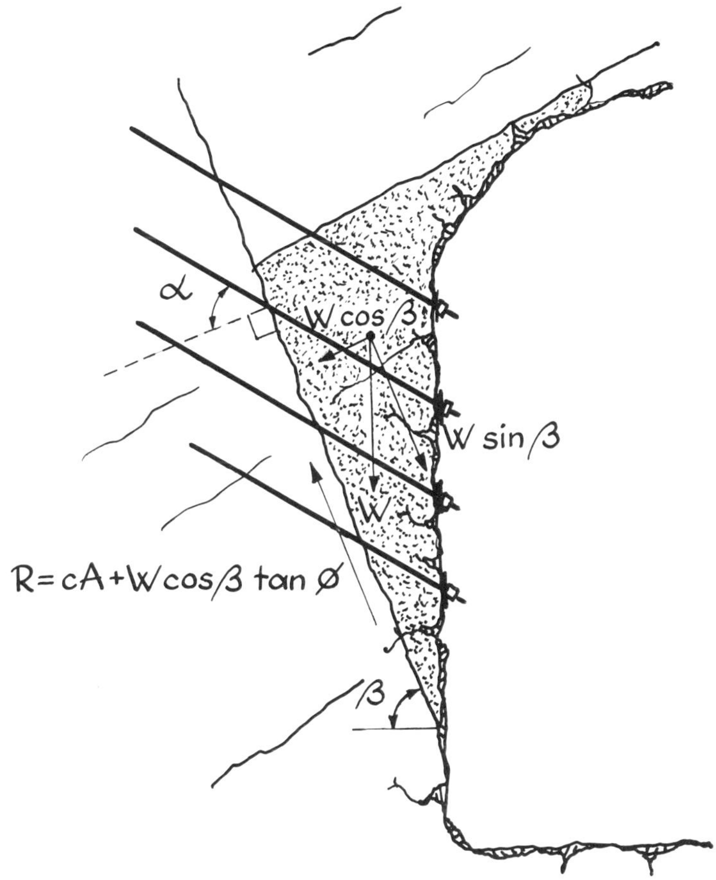

We have

$$N = \frac{W\ (f\sin\beta - \cos\beta\ \tan\phi) - cA}{B\ (\cos\alpha\ \tan\phi + f\ \sin\alpha)} \qquad (6)$$

where

N = number of rockbolts
W = weight of wedge including the weight of any rock that is supported by the wedge, (density of rock approximately 2.7 t/m^3)
f = safety factor. It is often found that $1.5 \leqslant f < 3$. The value of f depends on the damage which would result from sliding of the wedge and also on whether permanent or temporary reinforcement is to be considered
β = dip of the sliding surface
ϕ = angle of friction of the sliding surface
c = cohesive strength of the sliding surface
A = base area of the sliding surface
B = load bearing capacity of the bolt, see also section 5.1
α = angle between the plunge of the bolt and the normal to the sliding surface

Fig. 19 Reinforcement of a wedge of rock free to slide under its own weight.

In the above analysis, the angle of friction and the cohesive strength of the sliding surface, as well as the load bearing capacity of the bolt at a particular inclination to the sliding surface, are all properties that may be difficult to measure. If test data from a particular site are not available, the data must be obtained by reference to the appropriate literature.

Typical values of the angle of friction, ϕ for some common rock types are listed in Table 1, section 6.1.

The value on the cohesive strength may vary considerably. Its value depends on the type of joint filling material and its water content. A typical value of c for a clay filled joint in a wide range of rocks is less than 1. The contribution from the cohesive strength is often neglected, that is, the value of c is set equal to zero. Zero cohesive strength effectively implies a larger safety factor in the final design of the reinforcement system.

General guidelines on the estimation of the load bearing capacity of the rockbolt at a particular angle to the sliding surface can be obtained from section 5.1.

Simpler forms of wedge analysis may be performed manually, or with a programmable calculator. For major projects, it is desirable to use an interactive computer program in which the effect of variations in joint orientation and properties, groundwater pressures, surcharge loading, etc. on the reinforcement required may be calculated quickly.

The beam or slab concept for bedded rock

In horizontally-bedded sedimentary rocks, the roof is controlled by the bedding plane weakness. Thus, excavations in sedimentary rocks are usually formed with flat or haunched roofs, a structural beam or slab, using rockbolts may then be formed, PANEK (1964).

This concept may be extended to side walls, where high axial loading leads to potential buckling of intact vertical slabs or columns.

The beam or slab concept for bedded rock with an overlaying competent strata:

The beam or slab concept for bedded rock with overlaying competent strata is used in horizontally-bedded rocks, in which the structural behaviour is controlled by bedding plane weakness, but where these planes are overlayed by a competent strata.

The bolts required for the support of the bedding planes are anchored in the overlaying solid rock. The number of bolts required is calculated, using only the dead weight of the rock.

In this type of ground the competent overlaying strata is usually consistent with regard to its thickness and its location in the rock formation. This facilitates the design of the reinforcement.

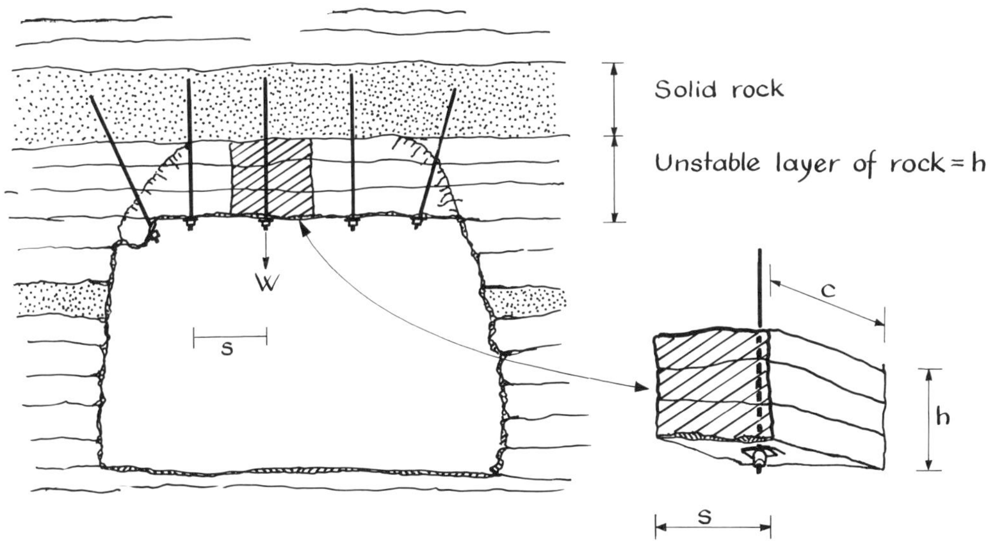

We have

$$W = f \times s \times c \times h \times \varrho \tag{7}$$

where

W = weight of rock to be supported by a single bolt
f = safety factor. It is often found that $1.5 \leqslant f < 3$. The value of f depends on the damage which would result from falling rock and also on whether permanent or temporary reinforcement is to be considered
s = bolt spacing, perpendicular to the axis of excavation
c = bolt spacing, along the axis of excavation
h = thickness of unstable layer of rock
ϱ = rock density, approximately 2.5 t/m^3

Fig. 20 Reinforcement of an unstable layer of horizontal bedding planes overlayed by solid rock.

The rockbolts should be slightly inclined at the abutments so that their action in shear is stronger, because of the presence of shear stresses from the slab action of the unstable layer of rock.

The beam or slab concept for bedded rock:

In horizontally-bedded laminated rock, tensioned bolts can be used to make the bedding planes interact and thereby increase the stability of the roof. The support of horizontally-bedded roofs in weak rock has been described by e.g. PANEK (1964) who has also formulated a practical design procedure for tensioned rockbolts under these conditions. This design procedure, based on the use of a special nomogram, includes the following steps, see Figs. 21 and 22;

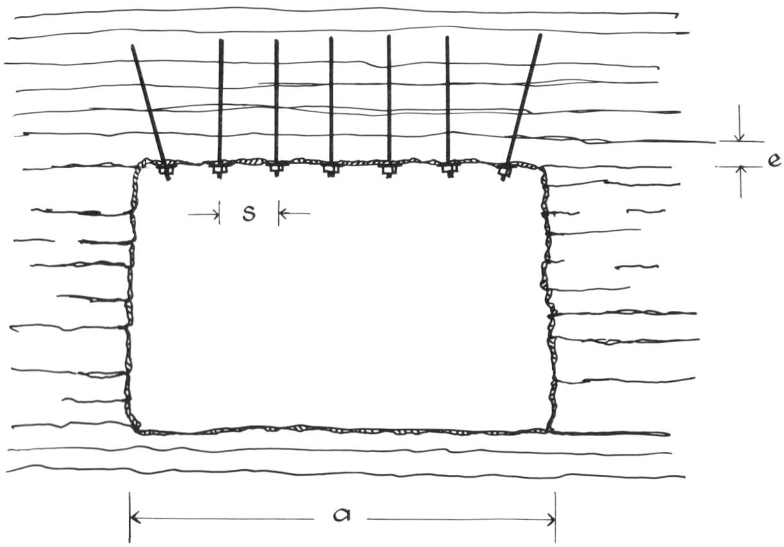

Fig. 21 Reinforcement of a horizontally-bedded roof in weak rock to form a beam or slab.

1. Estimate the mean thickness, e of the individual layers of rock. Choose a rockbolt length which permits the bolt to be anchored in a layer of sufficient thickness and strength for good anchoring of the bolt.
2. In pull-out tests, the correct level of tensioning is determined, (the tension applied to the bolts should be approximately 0.6 x the tensile strength of the bolt).
3. The number of bolts per set (in each row), perpendicular to the axis of the opening is estimated. If joints are present between the system of bolts, (joints that are likely to cause rock falls), the distance, s between the bolts should be adjusted accordingly, ($s \leqslant 3\ e$).
4. The spacing c between sets of bolts along the axis of the opening is selected. Normally, the spacing c is of the same order of magnitude as the distance s between the bolts.

5. + 6. By using the above information together with the known span of the opening in the chart, it is possible to determine the Reinforcement Factor (which should not be less than 2). The Reinforcement Factor is a measure of the

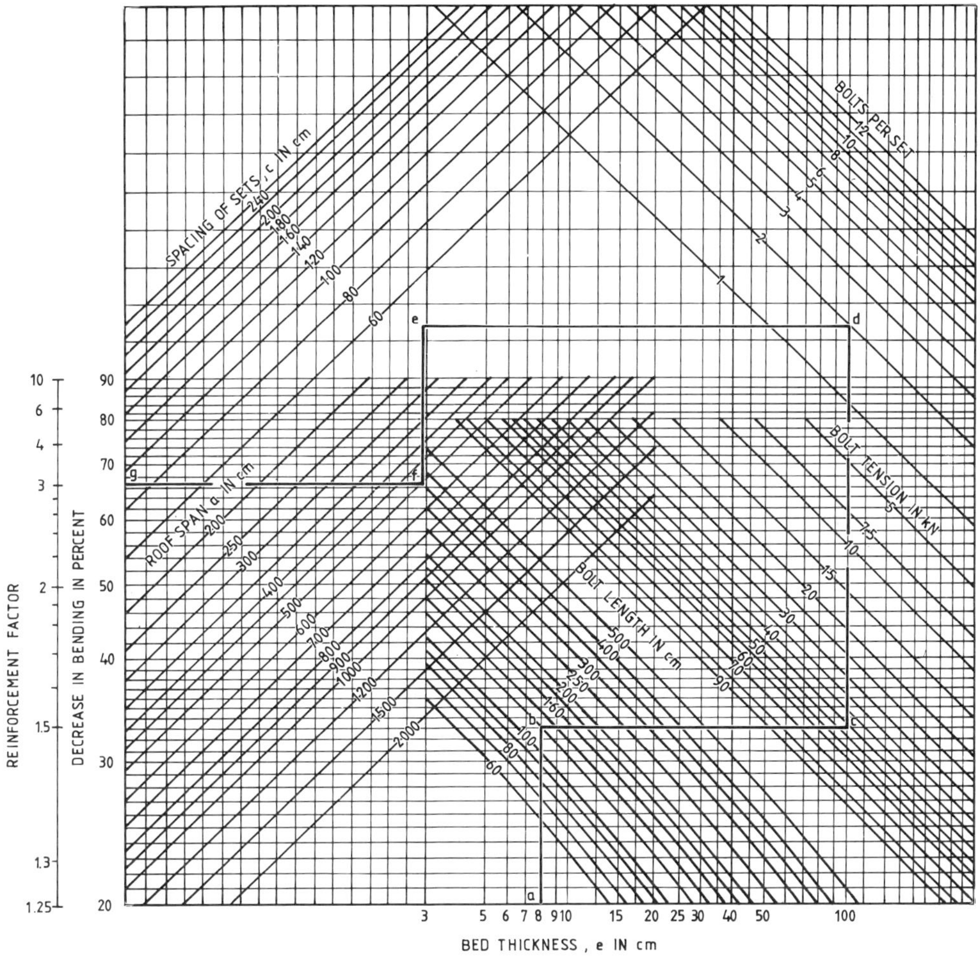

Fig. 22 Design chart for rockbolt reinforcement of horizontally-bedded weak rock. After PANEK (1964).

percentage decrease in the sag (bending) of the roof that is obtained from the rockbolting. A factor of 2 corresponds to a reduction in the sag by half, compared with an unsupported roof.

For example, assuming an opening with a span *a* of 5 metres (16 ft), a mean strata thickness *e* of 0.08 m (3 in), thickness of bolted strata, bolt length 1.2 m (4 ft), and bolt tension 4.5 tons (10000 lb), the construction a-b-c-d-e-f-g on the above chart shows that bolts spaced approximately (s = 1.7 m or 5.3 ft) across the opening (2 bolts per set, equally spaced) and 0.5 m (1.5 ft) apart along the opening (spacing of sets of bolts *c*) will result in a 67 % decrease in the maximum bending stress, which corresponds to a Reinforcement Factor of 3.

The rock arch concept

If an opening with a curved roof is excavated in a jointed rock mass (mosaic rock), a natural arch is formed at some depth into the rock, above the ceiling of the opening, see Fig. 23. The formation of the arch is a result of stress redistribution in the rock as the opening is formed. The rock in the arch is subjected primarily to compressive stresses. The distance between the ceiling and the lower boundary of the natural arch depends on the type of ground in which the opening is excavated and on the span of the opening. This phenomenon is called "the concept of natural arch development".

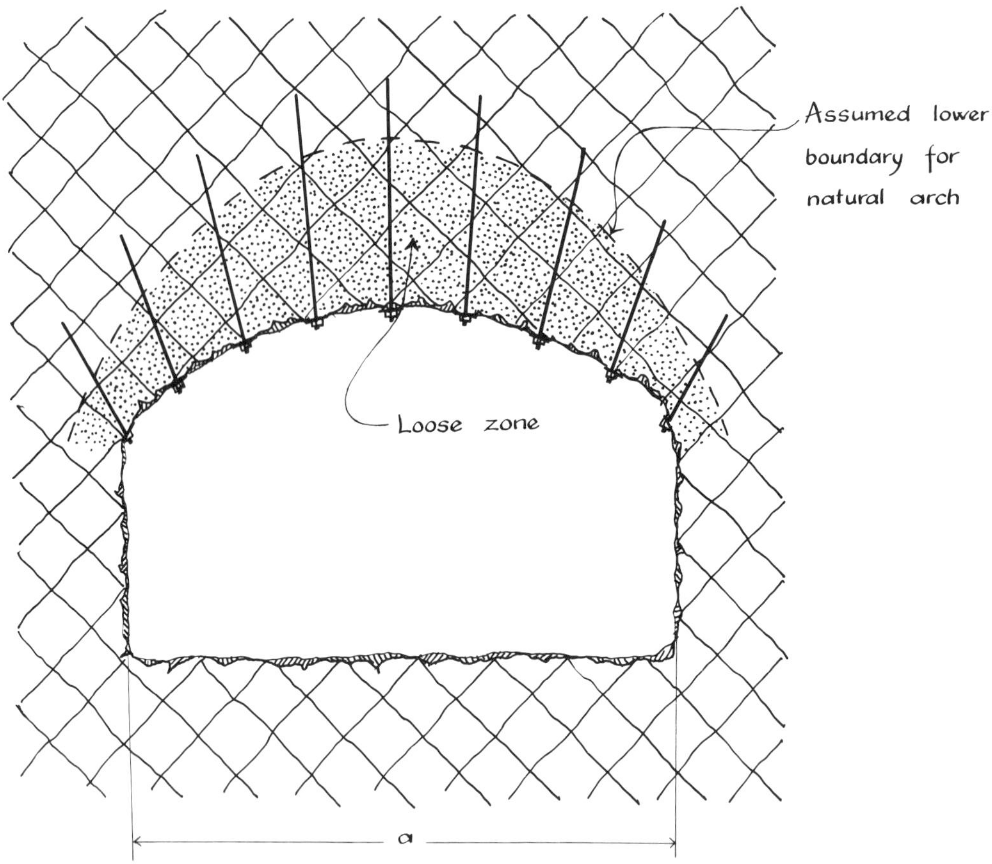

Fig. 23 The natural arch concept, which develops above the curved roof of an underground excavation in a moderately jointed rock. The "loose zone" of rock below the natural arch has been reinforced by untensioned rockbolts anchored in the natural arch.

To maintain the stability of the natural arch, the "loose zone" below the arch, as well as the walls of the opening must be stabilized. This can be achieved by reinforcing the rock with tensioned or untensioned rockbolts. Untensioned bolts, wherever they are used, must be anchored in the rock of the natural arch, see Fig. 23. The location of the lower boundary of the natural arch and hence the estimation of the length of the bolts is a problem. The Norwegian Institute for Rock Blasting Technique (IFF) has proposed a formula to estimate the length of the bolts in the central section of the opening which can be written in the form:

$$L = 1.40 + 0.184\,a \tag{8}$$

where a is the span of the opening measured in metres.

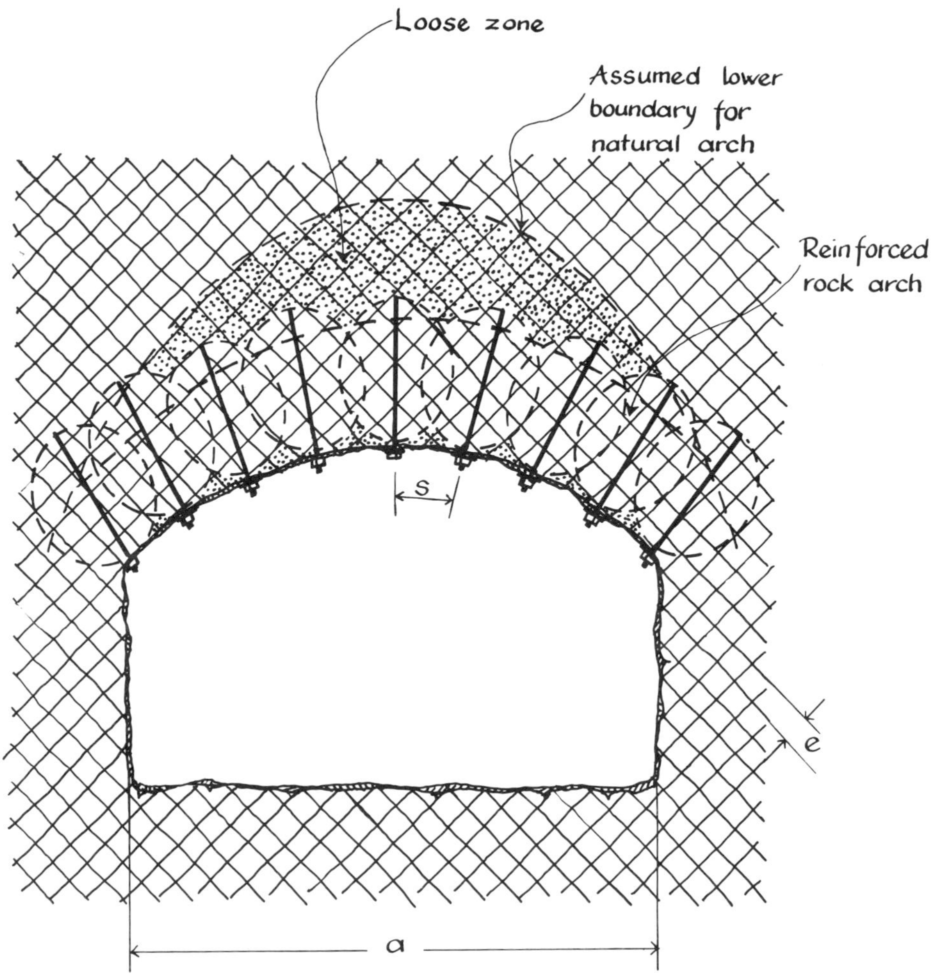

Fig. 24 The concept of a natural arch, developing above the curved roof of an underground excavation in heavily jointed rock. The "loose zone" of rock below the natural arch has been reinforced by tensioned rockbolts. This creates an artificial arch near the ceiling of the opening. Note the loose rock between the bolts near the ceiling of the opening. This rock must be supported by wire mesh or shotcrete.

The spacing between the bolts is determined by the joint density. Untensioned rockbolts are preferred in moderately jointed rocks, where the lower boundary of the natural arch is closer to the roof of the opening than is the case in heavily jointed rocks. Tensioned rockbolts, anchored in and used to reinforce the "loose zone" below the natural arch, form an artificial arch near the ceiling of the opening, see Fig. 24. This is of particular interest in less competent rock structures, where the lower boundary of the natural arch is further away from the roof of the opening. This concept is in contrast to the use of untensioned rockbolts anchored in the natural arch.

In a reinforced arch, an excavation is bounded by a zone under compression with high tangential/circumferential stresses. The rockbolts are used to increase both the shear resistance of the joints and the normal stress across joints. From experimental test results, it has been found that the ratio L/s of the length of the bolts, L, to the spacing between the bolts, s, should be close to 2. The spacing between the bolts s is dependent on the typical joint spacing e. The inequality $s \leqslant 3 \times e$ may be used as a limit for the bolt spacing s. Here, e is the joint spacing. The tension T, in the bolt should lie between the limits;

$$0.5\ B < T < 0.8\ B \qquad (9)$$

where

T = applied tension to the bolt
B = load bearing capacity of the bolt, (see also section 5.1)

The concept of reinforcing a curved roof excavation, to form a structurally competent arch, is fundamental to the design of rock reinforcement. It is used in most empirical design methods.

7.2.2 Empirical Assessments

Empirical assessments of rock reinforcement provide a useful supplement to any detailed analytical work. Empirical assessments can be very useful whenever adequate geotechnical data is unavailable for detailed structural analysis or whenever simplified analytical models are inapplicable. In the opinion of the author, provided sound engineering judgement is exercised for the conditions encountered, empirical assessments based on the use of the rock mass classification systems presented in Chapter 6, together with suitable monitoring systems, provide excellent guidelines for the design of appropriate support systems.

In connection with cable bolting, empirical assessments are often applied when long cables are used to reinforce large volumes of rock, e.g. a hanging-wall in mining. Structural data for the rock mass is very important, and detailed quantitative data on joint orientation and frequency forms an essential basis for design. However, a standard reinforcement pattern based on empirical results must commonly be adopted since detailed structural data is difficult to obtain for large volumes of rock. Three empirical assessment systems for rock reinforcement will be presented;

1. Empirical design recommendations according to the US Corps of Engineers.
2. Empirical assessments based on the Geomechanics Classification by Bieniawski.
3. Empirical assessments based on the Q-system by Barton et al.

Empirical design recommendations according to the U.S. Corps of Engineers

A system of simple recommendations for rockbolt reinforcement design has been formulated by the U.S. Corps of Engineers. The empirical rules, given in Table 5, are a summary of many important rock reinforcement case histories. It may be used as a guide for minimum reinforcement to be included in the preliminary design. The detailed design, based on further geological information obtained during the excavation stage, usually indicates the need for additional reinforcement. The recommendations have been widely used in practice.

Table 5
Typical empirical design recommendations. After U.S. Corps of Engineers (1980) and Douglas and Arthur (1983).

Parameter	*Empirical rule*
Minimum length and maximum spacing	
Minimum length	Greatest of: (a) 2 x bolt spacing (b) 3 x thickness of critical and potentially unstable rock blocks (Note 1) (c) For elements above the springline: spans <6 m: 0.5 x span spans between 18 and 30 m: 0.25 x span spans between 6 and 18 m: interpolate between 3 and 4.5 m (d) For elements below the springline: height <18 m: as (c) above height >18 m: 0.2 x height
Maximum spacing	Least of: (a) 0.5 x bolt length (b) 1.5 x width of critical and potentially unstable rock blocks (Note 1) (c) 2.0 m (Note 2)
Minimum spacing	0.9 to 1.2 m
Minimum average confining pressure	
Minimum average confining pressure at yield point of elements (Note 3)	Greatest of: (a) Above springline: *either* pressure = vertical rock load of 0.2 x opening width or 40 kN/m² (b) Below springline: *either* pressure = vertical rock load of 0.1 x opening height or 40 kN/m² (c) At intersections: 2 x confining pressure determined above (Note 4)

Notes:

1. Where joint spacing is close and span relatively large, the superposition of two reinforcement patterns may be appropriate (e.g. long heavy elements on wide centres to support the span, and shorter, lighter bolts on closer centres to stabilise the surface against ravelling).
2. Greater spacing than 2.0 m makes attachment of surface support elements (e.g. weldmesh or chain link mesh) difficult.
3. Assuming the elements behave in a ductile manner.
4. This reinforcement should be installed from the first opening excavated prior to forming the intersection. Stress concentrations are generally higher at intersections and rock blocks are free to move toward both openings.

Table 6
Geomechanics Classification guide for excavation and support in rock tunnels. After Bieniawski (1979)

SHAPE: HORSESHOE; WIDTH: 10 m; VERTICAL STRESS: BELOW 25 MPa; CONSTRUCTION: DRILLING AND BLASTING

Rock mass class	Excavation	Support		
		Rockbolts (20 mm, dia., fully bonded)*	Shotcrete	Steel sets
Very good rock I RMR: 81—100	Full face 3 m advance.	Generally no support required except for occasional spot bolting.		
Good rock II RMR: 61—80	Full face 1.0—1.5 m advance. Complete support 20 m from face.	Locally bolts in crown 3 m long, spaced 2,5 m with occasional wire mesh.	50 mm in crown where required.	None
Fair rock III RMR: 41—60	Top heading and bench 1.5—3 m advance in top heading. Commence support after each blast. Complete support 10 m from face.	Systematic bolts 4 m long, spaced 1.5 m—2 m in crown and walls with wire mesh in crown.	50—100 mm in crown and 30 mm in sides.	None
Poor rock IV RMR: 21—40	Top heading and bench 1.0—1.5 m advance in top heading. Install support concurrently with excavation 10 m from face.	Systematic bolts 4—5 m long, spaced 1—1.5 m in crown and walls with wire mesh.	100—150 mm in crown and 100 mm in sides.	Light to medium ribs spaced 1.5 m where required.
Very poor rock V RMR: <20	Multiple drifts. 0.5—1.5 m advance in top heading. Install support concurrently with excavation. Shotcrete as soon as possible after blasting.	Systematic bolts 5—6 m long, spaced 1—1.5 m in crown and walls with wire mesh. Bolt invert.	150—200 mm in crown, 150 mm in sides and 50 mm on face.	Medium to heavy ribs spaced 0.75 m with steel lagging and fore-poling if required. Close invert.

Empirical assessment of rock reinforcement based on the Geomechanics Classification

Based upon his Geomechanics Classification (section 6.2), Bieniawski has proposed a guide for the choice of support for underground excavations, which includes the use of rockbolts for rock reinforcement.

The support recommendations, reproduced in Table 6, apply to civil engineering tunnels of approximately 10 metres span, excavated by drilling and blasting methods at depths of less than 1000 metres (3300 ft) below surface.

Modification of the Geomechanics Classification support recommendations, for mining applications, has been proposed by Laubscher and Taylor (1976). They have suggested a number of modifications to both the rock mass classification system and the support recommendations. The modifications are based on the application of the rock mass classification to 50,000 metres of mining development and drill core. In the modifications proposed by Laubscher and Taylor, the original value of a particular parameter selected for incorporation in Table 2 (section 6.2) is changed by a percentage which is determined by the influence of weathering, field and induced stresses, stress redistribution from mining, excavation orientation and type (relative to geological structures), and the effect of blasting.

Hence, if an *RQD* value of 80 is obtained from drill core logging, and it is known that inferior blasting techniques were used for development, the *RQD* is reduced by 20%. This gives an adjusted rating of 64.

Guidelines for the calculation of adjustments are summarized below (after Laubscher and Taylor (1976)):

---

* **Supplementary note to Table 6 by** Stillborg

The rockbolts are referred to as "20 mm dia., fully bonded". This note implies that fully cement or resin grouted rockbolts are recommended. At the time when Bieniawski proposed his guide for excavation and support of rock tunnels, friction anchored rockbolts were not available. Friction anchored rockbolts have since become very popular, particularly in mining applications and more recently in tunnelling (the Swellex). Of the two friction anchored rockbolts, the Swellex acts like a fully bonded bolt, see section 7.1. The Swellex is simple and quick to install and gives immediate support action after installation. This makes it a very effective device for temporary support systems. In permanent support systems, when the length of the bolts must be varied and/or high capacity bolts must be used, fully grouted cable bolts become an interesting, inexpensive alternative to conventionally grouted rockbolts.

1. Weathering

Certain rock types weather readily on exposure. This must be considered in permanent support design. Three of the parameters listed in Table 2, the *RQD* value, the intact rock strength and rock joint conditions, are affected by weathering;

RQD: — the *RQD* can decrease by as much as 95 % as the rock weathers and the fracture density increases.

Intact rock strength: — may decrease by as much as 96 % if weathering occurs in rock micro-structures.

Condition of joints: — a decrease in rating of as much as 82 % is possible if weathering alters the joint wall rock and/or the joint filling.

2. Field and induced stresses

Field and induced stresses can affect the rock joint conditions. Under these stresses, both shear and dilation of the rock joint could occur, or alternatively compression of the rock joint could be maintained.

Condition of joints: — when the stress conditions are such that rock joint compression is maintained, the rating is increased by a maximum of 120 %. If the possibility of shear movement increases, decrease the rating by a maximum of 90 %. If the joints are open, they may be considered to be equivalent to joints with thin gouge filling; decrease the rating by a maximum of 76 %.

3. Stress changes

Large stress changes from mining operations, (for example when crown pillars are formed or in developments adjacent to caving areas or large stopes) will lead to a change in rock joint conditions as in the case of changes associated with field and induced stresses discussed above.
Condition of joints: — whenever rock joint compression is maintained by the stress changes, the rating is increased by a maximum of 120 %. Whenever the stress changes are likely to cause serious shear movement or joint opening, the rating is decreased by as much as 60 %.

4. Influence of strike and dip orientations

The shape, size and orientation of the excavation will when considered in connection with rock mass jointing, have an influence on the stability of the excavation. This is considered by BIENIAWSKI in his Geomechanics Classification, but LAUBSCHER and TAYLOR consider that a futher adjustment is necessary. They consider that the stability of an excavation in a jointed rock mass depends on the number of joints and excavation faces which are not vertical. Adjustments should also be made to the joint spacing rating for shear zones encountered in developments. The percentage adjustment to be applied depends on the shear zone orientations.

The following *Joint spacing rating* adjustment is suggested. This depends on the number of joints and excavation faces which are not vertical:

Percentage adjustment (depends on number of inclined excavation faces)

Number of joints	70 %	75 %	80 %	85 %	90 %
3	3		2		
4	4	3		2	
5	5	4	3	2	1
6	6		4	3	2,1

Proposed adjustments to the *Joint spacing rating* for shear zones in developments (depends on the orientation of the zones):

0°—15° = 76 %	15°—45° = 84 %	45°—75° = 92 %

5. Blasting effects

Blasting creates new fractures and generates movement on existing joints. The following reductions are proposed for *RQD* and *Condition of joints* ratings:

Technique	Adjustment
Boring	100 %
Smooth wall blasting	97 %
Good conventional blasting	94 %
Poor conventional blasting	80 %

6. Combined adjustments

In a number of situations, more then one adjustment will be applied to the Geomechanics Classification. According to LAUBSCHER and TAYLOR, the total adjustment applied should not exceed 50 %.

Based on the adjusted classification ratings and by considering typical mining support practices, LAUBSCHER und TAYLOR have proposed the support guide shown in Table 7.

Table 7
Modified Geomechanics Classification guide for support of mining excavations. After LAUBSCHER and TAYLOR (1976). See also supplementary note by STILLBORG, Page 71.

Adjusted ratings	Original rock mass ratings, *RMR*									
	90—100	80—90	70—80	60—70	50—60	40—50	30—40	20—30	10—20	0—10
60—100										
50—60		*a*	*a*	*a*	*a*					
40—50			*b*	*b*	*b*	*b*				
30—40				*c,d*	*c,d*	*c,d,e*	*d,e*			
20—30					*g*	*f,g*	*f,g,j*	*f,h,j*		
10—20						*i*	*i*	*h,i,j*	*h,j*	
0—10							*k*	*k*	*l*	*l*

a) Generally no support but locally joint intersections might require bolting.

b) Patterned grouted bolts at 1 m collar spacing.

c) Patterned grouted bolts at 0.75 m collar spacing.

d) Patterned grouted bolts at 1 m collar spacing and shotcrete 100 mm thick.

e) Patterned grouted bolts at 1 m collar spacing and massive concrete 300 mm thick and only used if stress changes not excessive.

f) Patterned grouted bolts at 0.75 m collar spacing and shotcrete 100 mm thick.

g) Patterned grouted bolts at 0.75 m collar spacing with mesh reinforced shotcrete 100 mm thick.

h) Massive concrete 450 mm thick with patterned grouted bolts at 1 m spacing if stress changes are not excessive.

i) Grouted bolts at 0.75 m collar spacing if reinforcing potential is present, and 100 mm reinforced shotcrete, and then yielding steel arches as a repair technique if stress changes are excessive.

j) Stabilise with rope cover support and massive concrete 450 mm thick if stress changes not excessive.

k) Stabilise with rope cover support followed by shotcrete to and including face if necessary, and then closely spaced yielding arches as a repair technique where stress changes are excessive.

l) Avoid development in this ground otherwise use support systems *j* or *k*.

Supplementary notes

I The original Geomechanics Classification as well as the adjusted ratings must be taken into account in assessing the support requirements.

II Bolts serve little purpose in highly jointed ground and should not be used as the sole support where the joint spacing rating is less than 6.

III The recommendations contained in Table 7 are applicable to mining operations with stress levels less than 30 MPa.

IV Large chambers should only be excavated in rock with adjusted total classification ratings of 50 or better.

Empirical assessment of rock reinforcement based on the *Q*-system

The *Q*-system for rock mass classification was reviewed in some detail in section 6.2 of this handbook. In Fig. 13 of the same section, the relationship between the maximum equivalent dimension D_e of an unsupported underground excavation, and the rock mass quality (*Q*) was illustrated. Before the review in section 6.2 is extended to deal with support recommendations, some further comments will be made related to permanently unsupported excavations.

In Fig. 25, the unsupported span for some 30 cases of permanently unsupported excavations, is plotted against the rock mass quality (*Q*), BARTON (1976). As pointed out by BARTON, and HOEK and BROWN (1980), there is no way of knowing how close these excavations are to failure. It could therefore be argued that Fig. 25 will always provide a conservative estimate of the unsupported excavation span. However, before deciding on a less conservative design, the consequences of possible stability problems should be carefully considered.

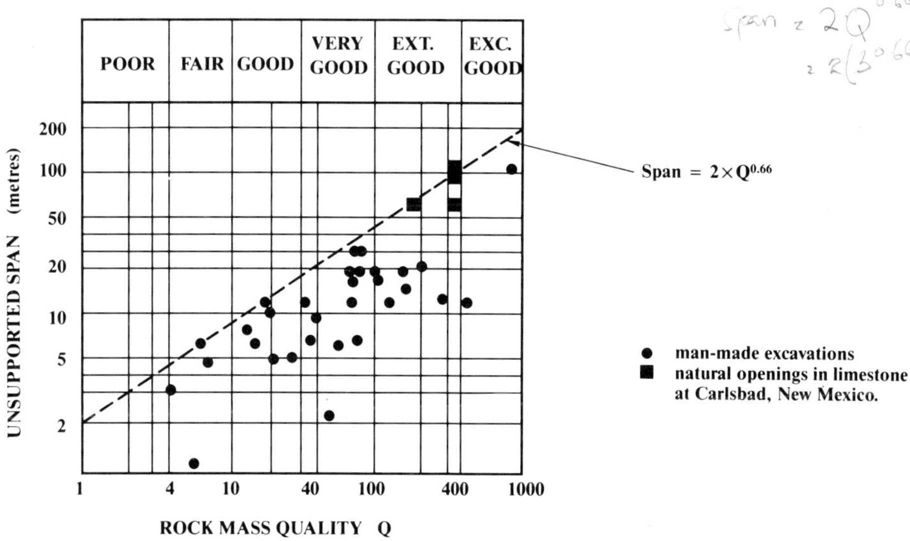

Fig. 25 Man-made and natural, unsupported excavations in different quality rock masses. After BARTON (1976).

In Fig. 26 the maximum unsupported spans for different Excavation Support Ratios, *ESR* have been plotted against the rock mass quality (Q).

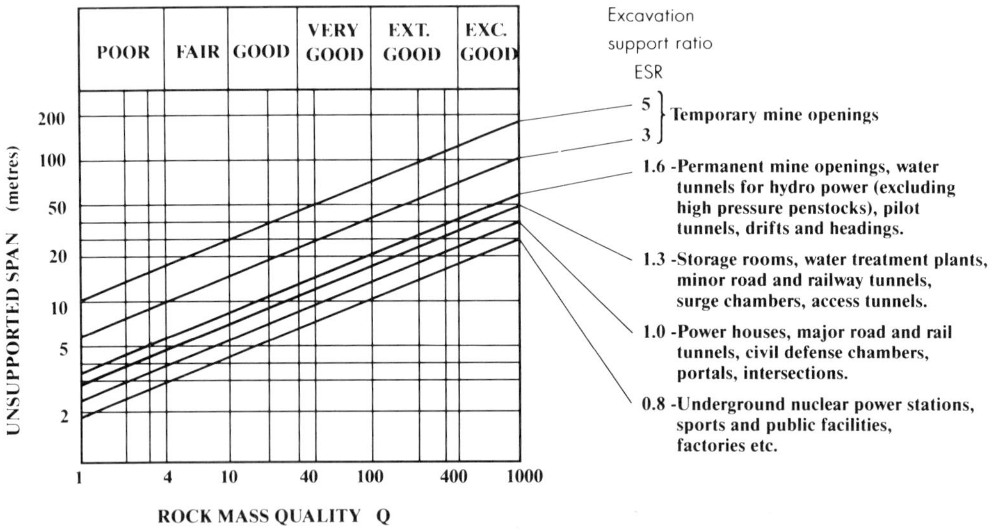

Fig. 26 Recommended maximum unsupported excavation spans for different rock mass quality (Q) and *ESR*-values. After BARTON (1976).

The equation which defines the lines plotted in Fig. 26 can be written:

$$\text{Span of opening} = 2 \times ESR \times Q^{0.4} \tag{10}$$

Alternatively, the critical value of Q for a given excavation span can be expressed in the form:

$$Q = \left(\frac{\text{Span}}{2 \times ESR}\right)^{2.5} \tag{11}$$

BARTON suggests that during the mapping of a tunnel route or the evaluation of an underground excavation site from drill cores, the knowledge of the critical value of Q will enable the engineer to identify those sections which can probably remain unsupported, and those which will require detailed consideration in terms of support requirements.

A limited percentage of the total number of the underground excavations developed are located in rock conditions and/or have a span such that they may remain permanently unsupported. For the majority of excavations, it is necessary to install some form of support to maintain stability of the surrounding rock mass. For excavations with a span exceeding the limits predicted by the above equations or

Fig. 26, Barton, Lien and Lunde (1974) have proposed 38 categories of support. These depends on the rock mass quality (*Q*) and the Excavation Support Ratio *ESR*, (see Fig. 27 and the detailed description in Table 8).

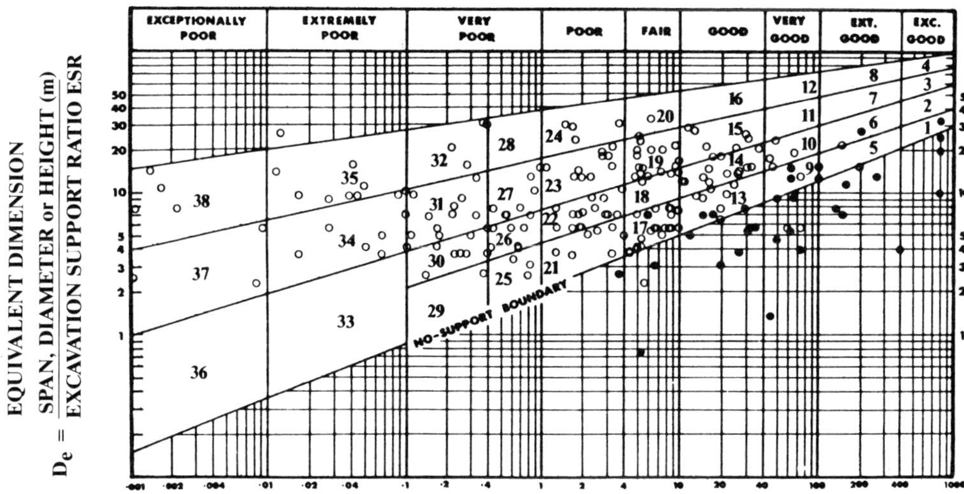

$$\text{ROCK MASS QUALITY} \quad Q = \frac{RQD}{J_n} \times \frac{J_r}{J_a} \times \frac{J_w}{SRF}$$

Fig. 27 Recommended support guide for different rock mass quality (*Q*) and *ESR*-values. After Barton et al. (1977). Support for categories 1 to 38 is listed in Table 8 (permanently unsupported openings).

2 X 1.6 × 3^0.4

= (6 / 2×1.6)^2.5

= 4.8

Q

Table 8
Suggested support measures for the 38 categories. After BARTON et al. (1977).

Support Measures for Rock Masses of "Exceptional", "Extremely Good", "Very Good" and "Good" Quality (*Q* range: 1000—10).

Support category	Conditional factors $\frac{RQD}{J_n}$	$\frac{J_r}{J_a}$	$\frac{Span}{ESR}$	Type of support	Note
1*	—	—	—	sb (utg)	—
2*	—	—	—	sb (utg)	—
3*	—	—	—	sb (utg)	—
4*	—	—	—	sb (utg)	—
5*	—	—	—	sb (utg)	—
6*	—	—	—	sb (utg)	—
7*	—	—	—	sb (utg)	—
8*	—	—	—	sb (utg)	—
9	⩾20	—	—	sb (utg)	—
	<20	—	—	B (utg) 2.5—3 m	—
10	⩾30	—	—	B (utg) 2—3 m	—
	<30	—	—	B (utg) 1.5—2 m	
				+ clm	—
11*	⩾30	—	—	B (tg) 2—3 m	—
	<30	—	—	B (tg) 1.5—2 m	
				+ clm	—
12*	⩾30	—	—	B (tg) 2—3 m	—
	<30	—	—	B (tg) 1.5—2 m	
				+ clm	—

Key to Support Tables:

sb = spot bolting
B = systematic bolting
(utg) = untensioned, grouted
(tg) = tensioned, (expanding shell type for competent rock masses, grouted post-tensioned in very poor quality rock masses; see Note XI)
S = shotcrete
(mr) = mesh reinforced
clm = chain link mesh
CCA = cast concrete arch
(sr) = steel reinforced

Bolt spacings are given in metres (m). Shotcrete, or cast concrete arch thickness is given in centimetres (cm).

Support category	Conditional factors $\frac{RQD}{J_n}$	$\frac{J_r}{J_a}$	$\frac{\text{Span}}{ESR}$	Type of support	Note
13	⩾10	⩾1.5	—	sb (utg)	I
	⩾10	<1.5	—	B (utg) 1.5—2 m	I
	<10	⩾1.5	—	B (utg) 1.5—2 m	I
	<10	<1.5	—	B (utg) 1.5—2 m + S 2—3 cm	I
14	⩾10	—	⩾15 m	B (tg) 1.5—2 m + clm	I, II
	<10	—	⩾15 m	B (tg) 1.5—2 m + S (mr) 5—10 cm	I, II
	—	—	<15 m	B (utg) 1.5—2 m + clm	I, III
15	>10	—	—	B (tg) 1.5—2 m + clm	I, II, IV
	⩽10	—	—	B (tg) 1.5—2 m + S(mr) 5—10 cm	I, II, IV
16* See note XII	>15	—	—	B (tg) 1.5—2 m + clm	I, V, VI
	⩽15	—	—	B (tg) 1.5—2 m + S (mr) 10—15 cm	I, V, VI

* Authors' estimates of support. Insufficient case records available for reliable estimation of support requirements.

Note: The type of support to be used in categories 1 to 8 will depend on the blasting technique. Smooth wall blasting and thorough barring-down may remove the need for support. Rough-wall blasting may result in the need for single applications of shotcrete, especially where the excavation height is >25 m. Future case records should differentiate categories 1 to 8.

Support Measures for Rock Masses of "Fair" and "Poor" Quality (Q range: 10-1).

Support category	Conditional factors $\frac{RQD}{J_n}$	$\frac{J_r}{J_a}$	$\frac{\text{Span}}{ESR}$	Type of support	Note
17	>30	—	—	sb (utg)	I
	(≥10, ≤30)	—	—	B (utg) 1—1.5 m	I
	<10	—	≥6 m	B (utg) 1—1.5 m + S 2—3 cm	I
	<10	—	<6 m	S 2—3 cm	I
18	>5	—	≥10 m	B (tg) 1—1.5 m + clm	I, III
	>5	—	<10 m	B (utg) 1—1.5 m + clm	I
	≤5	—	≥10 m	B (tg) 1—1.5 m + S 2—3 cm	I, III
	≤5	—	<10 m	B (utg) 1—1.5 m + S 2—3 cm	I
19	—	—	≥20 m	B (tg) 1—2 m + S (mr) 10—15 cm	I, II, IV
	—	—	<20 m	B (tg) 1—1.5 m + S (mr) 5—10 cm	I, II

Key to Support Tables:

sb = spot bolting
B = systematic bolting
(utg) = untensioned, grouted
(tg) = tensioned, (expanding shell type for competent rock masses, grouted post-tensioned in very poor quality rock masses; see Note XI)
S = shotcrete
(mr) = mesh reinforced
clm = chain link mesh
CCA = cast concrete arch
(sr) = steel reinforced

Bolt spacings are given in metres (m). Shotcrete, or cast concrete arch thickness is given in centimetres (cm).

Support category	Conditional factors $\frac{RQD}{J_n}$	$\frac{J_r}{J_a}$	$\frac{\text{Span}}{ESR}$	Type of support	Note
20* See note XII	—	—	⩾35 m	B (tg) 1—2 m + S (mr) 20—25 cm	I, V, VI
	—	—	<35 m	B (tg) 1—2 m + S (mr) 10—20 cm	I, II, IV
21	⩾12.5	⩽0.75	—	B (utg) 1 m + S 2—3 cm	I
	<12.5	⩽0.75	—	S 2.5—5 cm	I
	—	>0.75	—	B (utg) 1 m	I
22	(>10, <30)	>1.0	—	B (utg) 1 m + clm	I
	⩽10	>1.0	—	S 2.5—7.5 cm	I
	<30	⩽1.0	—	B (utg) 1 m + S (mr) 2.5—5 cm	I
	⩾30	—	—	B (utg) 1 m	I
23	—	—	⩾15 m	B (tg) 1—1.5 m + S (mr) 10—15 cm	I, II, IV, VII
	—	—	<15 m	B (utg) 1—1.5 m + S (mr) 5—10 cm	I
24* See note XII	—	—	⩾30 m	B (tg) 1—1.5 m + S (mr) 15—30 cm	I, V, VI
	—	—	<30 m	B (tg) 1—1.5 m + S (mr) 10—15 cm	I, II, IV

* Authors' estimates of support. Insufficient case records available for reliable estimation of support requirements.

Support Measures for Rock Masses of "Very poor" Quality (Q range: 1.0—0.1)

Support category	Conditional factors $\frac{RQD}{J_n}$	$\frac{J_r}{J_a}$	$\frac{Span}{ESR}$	Type of support	Note
25	>10	>0.5	—	B (utg) 1 m + mr or clm	I
	⩽10	>0.5	—	B (utg) 1 m + S (mr) 5 cm	I
	—	⩽0.5	—	B (tg) 1 m + S (mr) 5 cm	I
26	—	—	—	B (tg) 1 m + S (mr) 5—7.5 cm	VIII, X, XI
	—	—	—	B (utg) 1 m + S 2.5—5 cm	I, IX
27	—	—	⩾12 m	B (tg) 1 m + S (mr) 7.5—10 cm	I, IX
	—	—	<12 m	B (utg) 1 m + S (mr) 5—7.5 cm	I, IX
	—	—	>12 m	CCA 20—40 cm + B (tg) 1 m	VIII, X, XI
	—	—	<12 m	S (mr) 10—20 cm + B (tg) 1 m	VIII, X, XI

Key to Support Tables:

sb = spot bolting
B = systematic bolting
(utg) = untensioned, grouted
(tg) = tensioned, (expanding shell type for competent rock masses, grouted post-tensioned in very poor quality rock masses; see Note XI)
S = shotcrete
(mr) = mesh reinforced
clm = chain link mesh
CCA = cast concrete arch
(sr) = steel reinforced

Bolt spacings are given in metres (m). Shotcrete, or cast concrete arch thickness is given in centimetres (cm).

Support category	Conditional factors $\frac{RQD}{J_n}$	$\frac{J_r}{J_a}$	$\frac{\text{Span}}{ESR}$	Type of support	Note
28*	—	—	⩾30 m	B (tg) 1 m + S (mr) 30—40 cm	I, IV, V, IX
	—	—	(⩾20 m, <30 m)	B (tg) 1 m + S (mr) 20—30 cm	I, II, IV, IX
See note XII	—	—	<20 m	B (tg) 1 m + S (mr) 15—20 cm	I, II, IX
	—	—	—	CCA (sr) 30—100 cm + B (tg) 1 m	IV, VIII, X, XI
29*	>5	>0.25	—	B (utg) 1 m + S 2—3 cm	—
	⩽5	>0.25	—	B (utg) 1 m + S (mr) 5 cm	—
	—	⩽0.25	—	B (tg) 1 m + S (mr) 5 cm	—
30	⩾5	—	—	B (tg) 1 m + S 2.5—5 cm	IX
	<5	—	—	S (mr) 5—7.5 cm	IX
	—	—	—	B (tg) 1 m + S (mr) 5—7.5 cm	VIII, X, XI
31	>4	—	—	B (tg) 1 m + S (mr) 5—12.5 cm	IX
	⩽4, ⩾1.5	—	—	S (mr) 7.5—25 cm	IX
	<1.5	—	—	CCA 20—40 cm + B (tg) 1 m	IX
	—	—	—	CCA (sr) 30—50 cm + B (tg) 1 m	VII, X, XI
32	—	—	⩾20 m	B (tg) 1 m + S (mr) 40—60 cm	II, IV, IX
See note XII	—	—	<20 m	B (tg) 1 m + S (mr) 20—40 cm	III, IV, IX
	—	—	—	CCA (sr) 40—120 cm + B (tg) 1 m	IV, VIII, X, XI

* Authors' estimates of support. Insufficient case records available for confident prediction of support requirements.

Support Measures for Rock Masses of "Extremely Poor" and "Exceptionally Poor" Quality (Q range: 0.1—0.001)

Support category	Conditional factors $\frac{RQD}{J_n}$	$\frac{J_r}{J_a}$	$\frac{\text{Span}}{ESR}$	Type of support	Note
33*	≥2	—	—	B (tg) 1 m	IX
				+ S (mr) 2.5—5 cm	
	<2	—	—	S (mr) 5—10 cm	IX
	—	—	—	S (mr) 7.5—15 cm	VIII, X
34	≥2	≥0.25	—	B (tg) 1 m	IX
				+ S (mr) 5—7.5 cm	
	<2	≥0.25	—	S (mr) 7.5—15 cm	IX
	—	<0.25	—	S (mr) 15—25 cm	IX
	—	—	—	CCA (sr) 20—60 cm	VIII, X
				+ B (tg) 1 m	XI
35	—	—	≥15 m	B (tg) 1 m	II, IX
				+ S (mr) 30—100 cm	
See	—	—	≥15 m	CCA (sr) 60—200 cm	VIII, X,
note				+ B (tg) 1 m	XI, II
XII	—	—	<15 m	B (tg) 1 m	IX, III
				+ S (mr) 20—75 cm	
	—	—	<15 m	CCA (sr) 40—150 cm	VIII, X,
				+ B (tg) 1 m	XI, III
36*	—	—	—	S (mr) 10—20 cm	IX
	—	—	—	S (mr) 10—20 cm	VIII, X,
				+ B (tg) 0.5—1.0 m	XI
37	—	—	—	S (mr) 20—60 cm	IX
	—	—	—	S (mr) 20—60 cm	VIII, X,
				+ B (tg) 0.5—1.0 m	XI
38	—	—	≥10 m	CCA (sr) 100—300 cm	IX
	—	—	≥10 m	CCA (sr) 100—300 cm	VIII, X,
See				+ B (tg) 1 m	II, XI
note	—	—	<10 m	S (mr) 70—200 cm	IX
XIII	—	—	<10 m	S (mr) 70—200 cm	VIII, X,
				+ B (tg) 1 m	III, XI

* Authors' estimates of support. Insufficient case records available for confident prediction of support requirements.

Supplementary notes by Barton, Lien *and* Lunde

I. For cases of heavy bursting or "popping", tensioned bolts with enlarged bearing plates often used, with spacing of about 1 m (occasionally down to 0.8 m). Final support when "popping" activity ceases.

II. Several bolt lengths often used in same excavation, i.e. 3, 5 and 7 m.

III. Several bolt lengths often used in same excavation, i.e. 2, 3 and 4 m.

IV. Tensioned cable anchors often used to supplement bolt support pressures. Typical spacing 2—4 m.

V. Several bolt lengths often used in same excavation, i.e. 6, 8 and 10 m.

VI. Tensioned cable anchors often used to supplement bolt support pressures. Typical spacing 4—6 m.

VII. Several of the older generation power stations in this category employ systematic or spot bolting with areas of chain link mesh, and a free span concrete arch roof (25—40 cm) as permanent support.

VIII. Cases involving swelling, for instance montmorillonite clay (with access of water). Room for expansion behind the support is used in cases of heavy swelling. Drainage measures are used where possible.

IX. Cases not involving swelling clay or squeezing rock.

X. Cases involving squeezing rock. Heavy rigid support is generally used as permanent support.

XI. According to the authors' experience, in cases of swelling or squeezing, the temporary support required before concrete (or shotcrete) arches are formed may consist of bolting (tensioned shell-expansion type) if the value of RQD/J_n is sufficiently high (i.e. >1.5), possibly combined with shotcrete. If the rock mass is very heavily jointed or crushed (i.e. RQD/J_n <1.5, for example a "sugar cube" shear zone in quartzite), then the temporary support may consist of up to several applications of shotcrete. Systematic bolting (tensioned) may be added after casting the concrete, but it may not be effective when RQD/J_n <1.5 or when a lot of clay is present, unless the bolts are grouted before tensioning. A sufficient length of anchored bolt might also be obtained using quick setting resin anchors in these extremely poor quality rock-masses. Serious occurrences of swelling and/or squeezing rock may require that the concrete arches are taken right up to the face, possibly using a shield as temporary shuttering. Temporary support of the working face may also be required in these cases.

XII. For reasons of safety the multiple drift method will often be needed during excavation and supporting of roof arch. Categories 16, 20, 24, 28, 32, 35 (SPAN/ESR >15 m only).

XIII. Multiple drift method usually needed during excavation and support of arch, walls and floor in cases of heavy squeezing. Category 38 (SPAN/ESR > 10 m only).

Supplementary notes by Hoek *and* Brown *(1980)*

A. Chainlink mesh is sometimes used to catch small pieces of rock which can become loose with time. It should be attached to the rock at intervals of between 1 and 1.5 m and short grouted pins can be used between bolts. Galvanised chainlink mesh should be used where it is intended to be permanent, e.g. in an underground powerhouse.

B. Weldmesh, consisting of steel wires set on a square pattern and welded at each intersection, should be used for the reinforcement of shotcrete since it allows easy access of the shotcrete to the rock. Chainlink mesh should never be used for this purpose since the shotcrete cannot penetrate all the spaces between the wires and air pockets are formed with consequent rusting of the wire. When choosing weldmesh, it is important that the mesh can be handled by one or two men working from the top of a high-lift vehicle and hence the mesh should not be too heavy. Typically, 4.2 mm wires set at 100 mm intervals (designated 100 x 100 x 4.2 weldmesh) are used for reinforcing shotcrete.

C. In poorer quality rock, the use of untensioned grouted dowels as recommended by BARTON, LIEN and *Lunde* depends upon immediate installation of these reinforcing elements behind the face. This depends upon integrating the support drilling and installation into the drill-blast-muck cycle and many non-Scandinavian contractors are not prepared to consider this system. When it is impossible to ensure that untensioned grouted dowels are going to be installed immediately behind the face, consideration should be given to using tensioned rockbolts which can be grouted at a later stage. This ensures that support is available during the critical excavation stage.

D. Many contractors would consider that a 200 mm thick cast concrete arch is too difficult to construct because there is not enough room between the shutter and the surrounding rock to permit easy access for pouring concrete and placing vibrators. The US Army Corps of Engineers (1978) suggests 10 inches (254 mm) as a normal minimum while some contractors prefer 300 mm.

E. BARTON, LIEN and LUNDE suggest shotcrete thicknesses of up to 2 m. This would require many separate applications and many contractors would regard shotcrete thicknesses of this magnitude as both impractical and uneconomic, preferring to cast concrete arches instead. A strong argument in favour of shotcrete is that it can be placed very close to the face and hence can be used to provide early support in poor quality rock masses. Many contractors would argue that a 50 to 100 mm layer is generally sufficient for this purpose, particularly when used in conjunction with tensioned rockbolts as indicated by BARTON, LIEN and LUNDE, and that the placing of a cast concrete lining at a later stage would be a more effective way to tackle the problem. Obviously, the final choice will depend upon the unit rates for concreting and shotcreting offered by the contractor and, if shotcrete is cheaper, upon a practical demonstration by the contractor that he can actually place shotcrete to this thickness.
In North America, the use of concrete or shotcrete linings of up to 2 m thick would be considered unusual and a combination of heavy steel stets and concrete would normally be used to achieve the high support pressures required in very poor ground.

Supplementary note by STILLBORG

Untensioned, grouted rockbolts are recommended in several support categories. At the time when BARTON et al. proposed their guide for support measures the friction anchored rockbolts were not yet available. The note under Table 6 in connection with BIENIAWSKI's guide for excavation and support in rock tunnels is therefore equally applicable here.

Figures 28 A-C show some examples of *Q*-system mapping. The rock mass conditions are described by the six independent parameters *RQD*, J_n etc., (section 6.2). Both the existing temporary support (if any) and the recommended permanent support are recorded.

In Figs. 28 A-C the recommended support was selected with the help of Fig. 27 and Table 8. However, this selection may be modified in special circumstances.

One should bear in mind that neither the *Q*-system nor the Geomechanics Classification adequately covers the situation described in the previous section, when the stability of isolated blocks or wedges of rock were considered. Whenever an underground opening is excavated, and no systematic reinforcement is considered necessary, on the basis of Table 6 or Fig. 27, the need for rock bolting to stabilize blocks or wedges of rock must still be recognized.

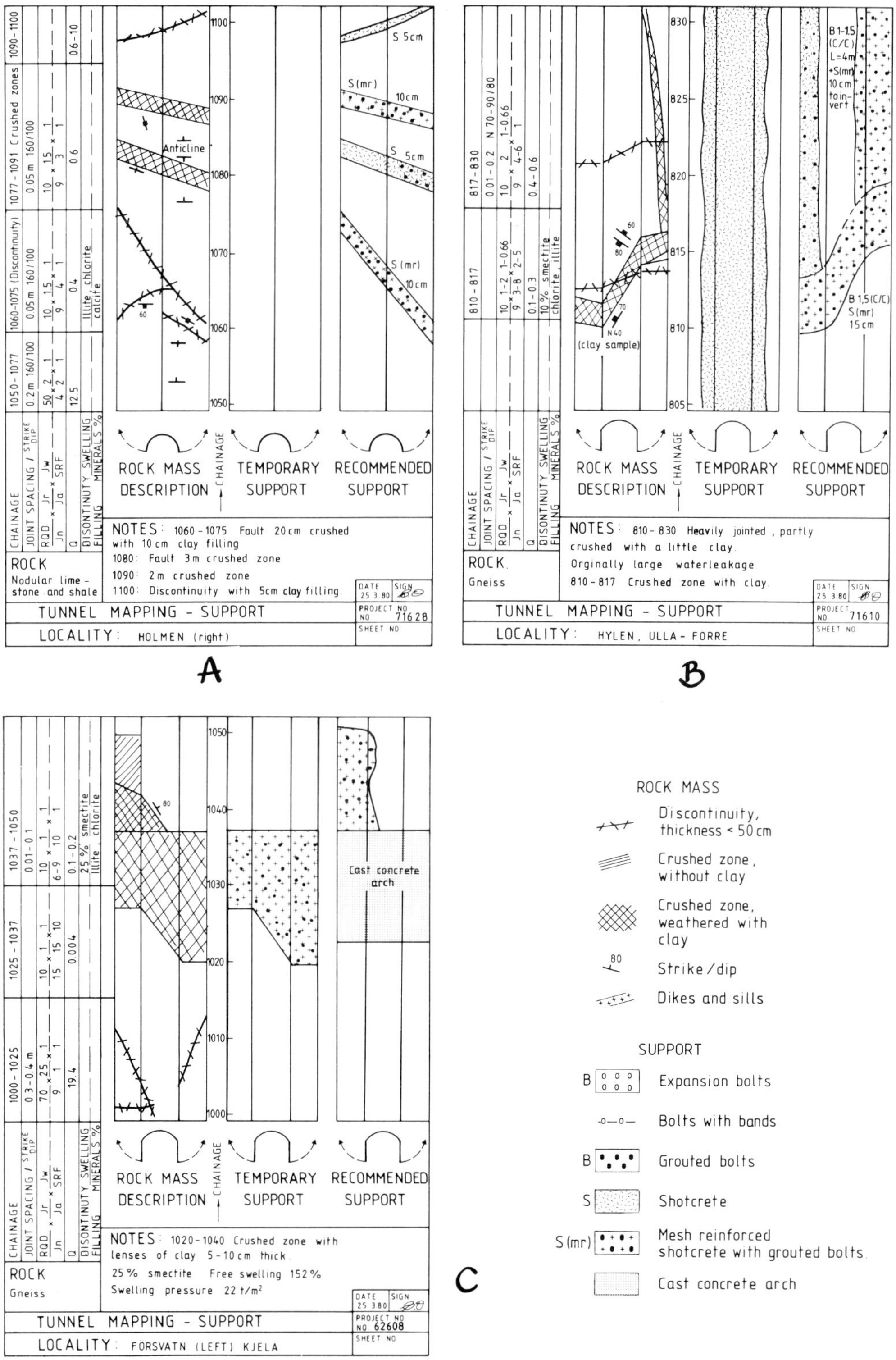

Fig. 28A-C Examples of tunnel mapping according to the *Q*-system. After BARTON et al. (1980).

A rock reinforcement drawing must be presented in such a way that all necessary information is included. Also, the drawing should be easily read and understood. A rock reinforcement design, however well suited to the conditions encountered, is of limited use unless it is applied to the intentions of the designer.

The following information should be included in a rock reinforcement drawing:

Example

Rockbolt reinforcement specifications

Type of rockbolt	Swellex
Bolt tension	—
Bolt length	2.4 metres
Borehole diameter	38 mm
Distance between bolts *b*	1.5 metres
Distance between rows *r*	2.0 metres
Direction and location of bolts	see reinforcement drawing

Rockbolt reinforcement drawing

Example

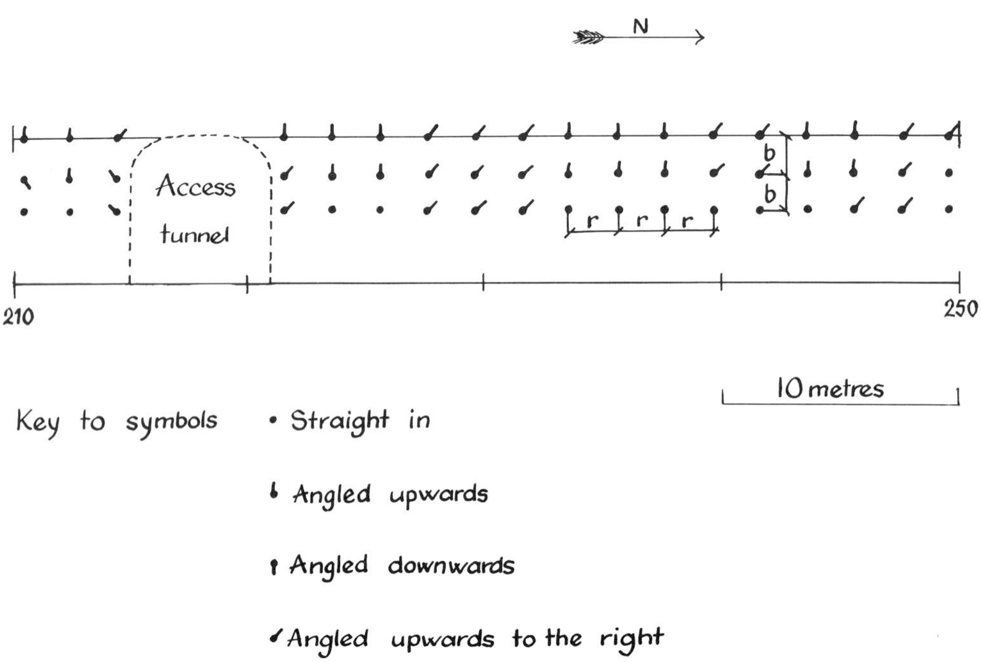

Fig. 29 Rockbolt reinforcement drawing, showing left hand side of tunnel looking northwards. (The same rockbolt location and direction symbols are painted at the tunnel wall).

7.2.3 Numerical Models

Numerical modelling, such as finite element techniques, (FEM), Haas et al. (1979) and Beer et al. (1985) or boundary element techniques, (BEM), Coulthard et al. (1983) and Pande (1985) can be used to study rock reinforcement. In these models, regions of rock containing joints and rockbolts are represented by special elements. The deformation of these elements, under specified boundary conditions, is analysed by a computer.

One advantage of a numerical model is the ease with which the deformation of a simulated engineering structure can be calculated. The engineer can then compare measured and calculated deformations and hence progressively improve the design of the rock reinforcement, for example. There are, however, a number of problems involved in the application of any numerical technique to the design of rock reinforcement. For example, uncertainty exists in the geotechnical input data and also in the correct modelling of the reinforcing elements, etc.

Undoubtedly, current numerical models are the forerunners of a vast number of future numerical techniques. These techniques will form a very valuable complement to other rock reinforcement design principles. Here, the author would like to quote Hoek and Brown (1980):

"The authors have no objection to these developments provided that both the numerical analysts and the users of these methods bear in mind the real world in which elegant theoretical models can be rendered invalid by the presence of a clay filled fault, the careless design of a blast resulting in a poor excavation profile, or the failure of a grout pump during a concrete lining operation resulting in voids behind the lining. The essential ingredient of any successful rock support programme is the ability of the underground excavation engineer to adapt to the actual conditions encountered in the field. It is important that this engineer should start out with a clear understanding of what he or she is trying to achieve and a sound idea, based upon theoretical studies or on previous experience, of what support options are available. It is even more important that this engineer should be able to adapt these options to the constantly changing conditions encountered underground. The rigid adherence to a single support design, however elegant, does not produce an effective or economical underground excavation."

Numerous instruments are available for underground use. It is beyond the scope of this handbook to deal with all of these instruments. A few of the most important instruments and measuring techniques will be reviewed. However, the reader is left to follow up details in catalogues and manuals from instrument manufacturers.

Before excavation

Detailed description of instrumentation and equipment for the collection of structural geological data can be found in a number of books, for example, Hoek and Brown (1980), and for rock strength and deformation characteristics, for example, Lama and Vutukuri (1978).

Rock stress measurements: — one of the most important design criteria for rock reinforcement is the in-situ rock stress conditions. With a few exceptions a detailed knowledge of the in-situ stress state is, however, unnecessary. In general, the most relevant information for the determination of underground excavation stability is, in order of importance, as follows: The principal stress directions. The relation between principal stress directions. The absolute magnitudes of stresses. With this in mind, any small differences between possible stress measurement methods, which may be thought to affect the choice of method in a particular situation, become less important. More important considerations concern the availability of instrumentation and experienced people. The ultimate success of a field exercise depends on practical details. This has been demonstrated by Stillborg and Leijon (1982) for example.

The fully three-dimensional stress field can be determined from a single borehole, using an over-coring technique, for example, the CSIRO Hollow Inclusion stress gauge. This Cell is supplied with nine or twelve strain gauges fully encapsulated in epoxy resin. After the Cell has been glued in the borehole using epoxy resin, strain readings can be taken from the Cell while over-coring is performed. This technique has been used extensively and very reliable results have been obtained, Stillborg and Leijon (1982).

Groundwater measurements: — this is an important practical problem in underground excavations. Rock mass stability can be severely affected by the presence of excessive groundwater. Many groundwater problems are difficult to quantify with any degree of precision. However, instruments and equipment are available to measure the flow rate of groundwater at different levels of a borehole. From measurements in several boreholes, it is possible to estimate rock mass permeability and groundwater pressure distribution in a rock mass. A comprehensive discussion on the permeability of jointed rocks can be found in Rissler (1978).

During excavation

During excavation, measurements should be performed either to check the validity of the design of the rockbolt system, or to permit the completion of on-going design work. In addition, these measurements should provide information which will ensure that appropriate measures are taken to prevent conditions from deteriorating to a point where they become very expensive or impossible to solve.

It is important that simple and rugged instruments are employed for measurements in the construction phase. They should be easily read, and should not interfere with on-going excavation activities. In the opinion of the author, deformation measurements provide the most effective means of monitoring the behaviour, not only of a rockbolt system but also of the rock mass behaviour in general. It is preferable that measurements on the rockbolts themselves, on the opening contour, as well as in the surrounding rock mass, are taken simultaneously.

Measurements on rockbolts: — with a torque wrench it is possible to obtain an approximate relationship between the torque and the tension in the bolt for mechanically anchored rockbolts. Despite the fact that a rusty bolthead will distort the measurements this technique is still very useful in the detection of inactive bolts. This type of feed-back can be used to determine the effectiveness of bolt design in various mine workings, and to establish new design specifications for bolting.

For the Split Set rockbolts, detachment of the bolt ring and consequent loss of the face plate contact at the hole collar can be used as an early warning that additional bolts are required, SCOTT (1984).

A slightly modified version of the Swellex bolt can be used to monitor the deformation of the rock. Referring to Figure 30, portions of the weld and lower bushing have been machined away. This provides surfaces on which measurements of the displacement of the lower bushing, relative to the main tube can be made. A thin steel tube fitted over the lower portion of the bolt prevents all but the last couple of feet, from expanding in the hole. The installation of these monitoring units at regular intervals within the rockbolt pattern provides a rapid and very useful system for the determination of rock mass stability.

Convergence measurements: — these are normally performed by means of a tape or a rod extensometer between measuring marks at the walls and roof of the opening. Figure 31 illustrates a typical convergence measuring array for a tunnel. The array includes two 8 metre long borehole rod extensometers. With the borehole extensometers, the directions of the deformations of the walls and roof of the tunnel can be determined. Even a piece of blasting wire which has been attached to a roof bolt and at the free end has a dead weight, can be used as a crude yet useful instrument to give warning of roof convergence. The change of distance between the dead weight and a block of rock can be measured by a folding rule, see Fig. 31.

Borehole extensometers: — are used to measure deformations in the rock mass surrounding an underground excavation. The rod extensometer is the most versatile of the different extensometers available. It consists of one or several rods each connected to an anchor and sheathed in a plastic tubing, see Fig. 32 for details.

The read-out system can be manual using a dial gauge or a depth micrometer, but remote read-out systems are also available. The accuracy of a rod extensometer is

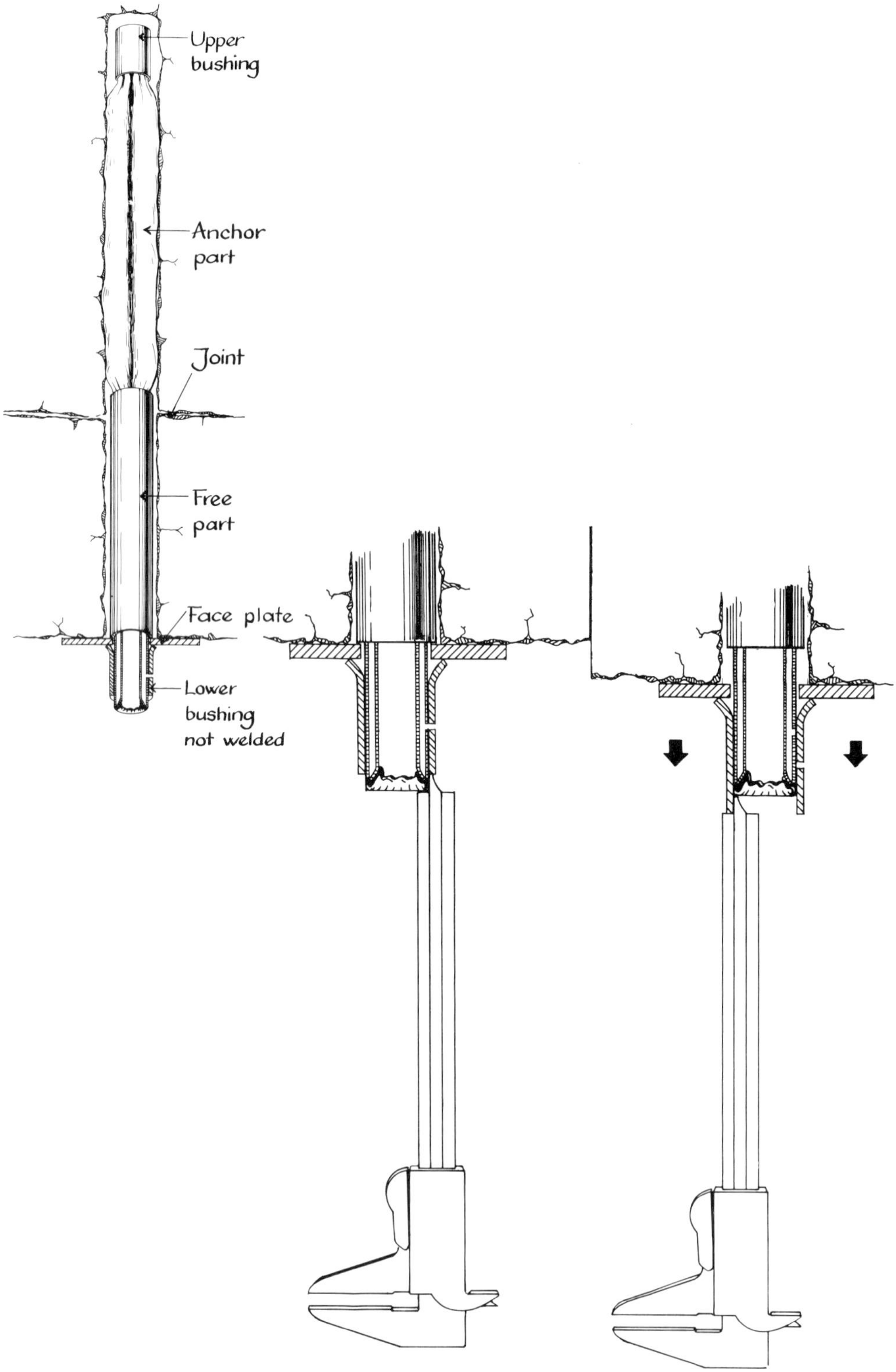

Fig. 30 The Swellex rock movement monitoring bolt. After MILLER (1984).

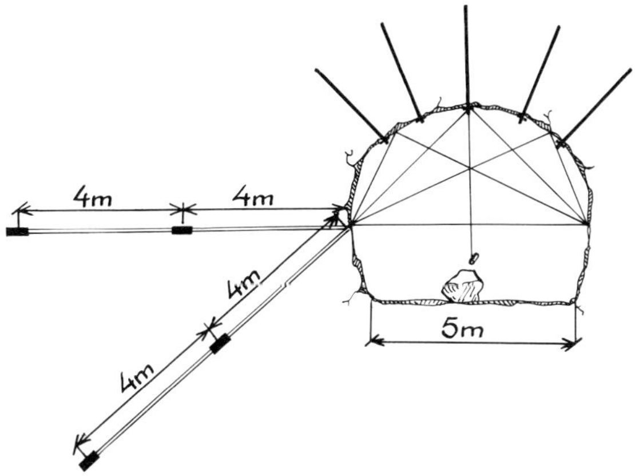

Fig. 31 Typical convergence measuring array including two borehole rod extensometers, for a bolted tunnel. A simple roof convergence measuring arrangement is also shown.

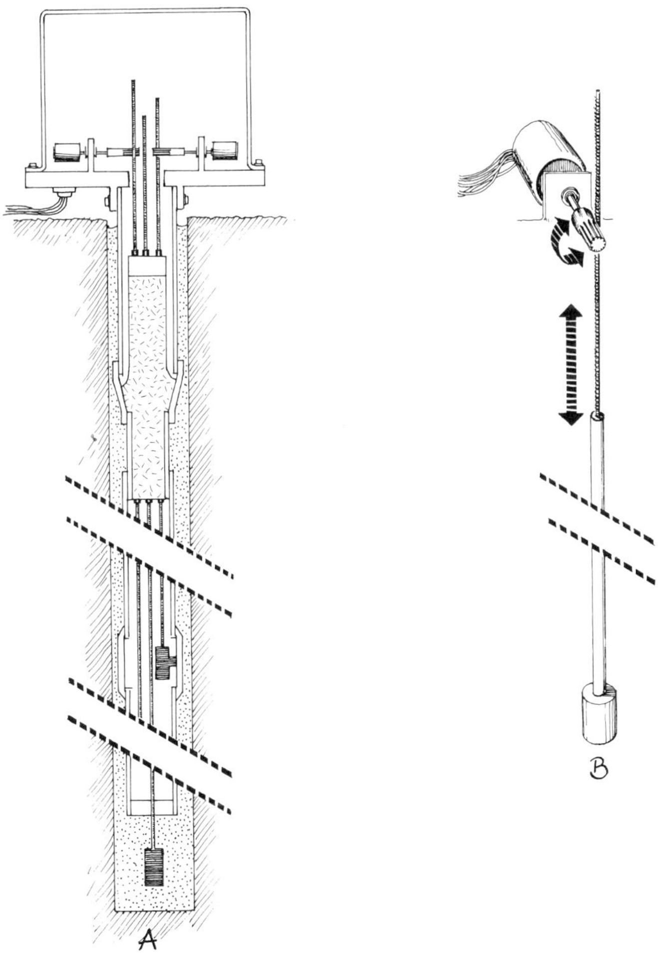

Fig. 32 Typical construction of a remote read multiple point borehole rod extensometer. A. Complete instrument. B. Details of potentiometer read-out system. After Stillborg (1983).

typically ± 0.1 mm. The readings from the extensometer should be presented graphically to facilitate interpretation of the results. Figure 33 gives four examples of the typical response from a one-anchor point extensometer. Comments are given on the respective graphs, with suggested changes in the reading intervals based on the shape of the graph.

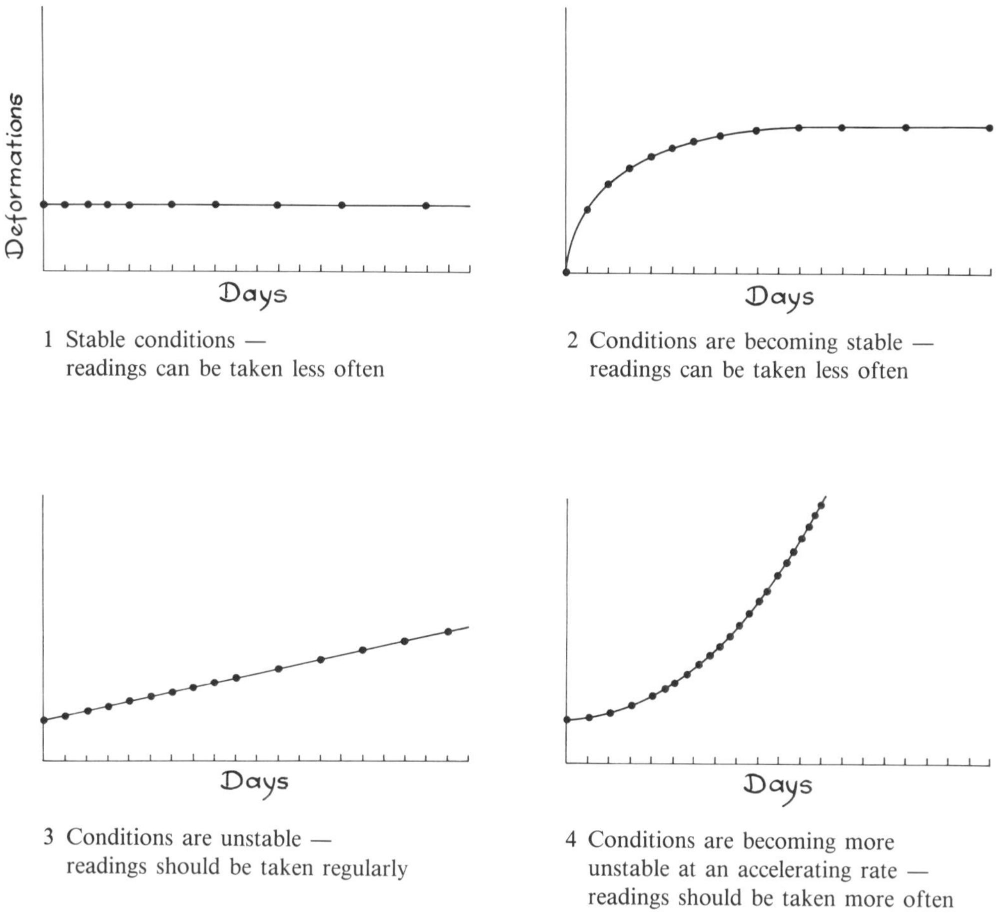

1 Stable conditions — readings can be taken less often

2 Conditions are becoming stable — readings can be taken less often

3 Conditions are unstable — readings should be taken regularly

4 Conditions are becoming more unstable at an accelerating rate — readings should be taken more often

Fig. 33 Examples of typical readings from a one-anchor point extensometer. In 3 and 4 one should also look for other signs of instability in the rock mass, for the appropriate action.

After excavation

The monitoring techniques after completion of the excavation are identical to those used during the excavation. However, the time interval between readings can be extended. In underground mining, monitoring of rock mass behaviour after the completion of the excavation is of particular importance, because of continuous changes in the stress state as a result of advanced mining.

Chapter 9

Cost of Rock Bolting

In any rock excavation project, it is important to identify the costs of the rock reinforcement. The information can then be used as follows:

— Evaluate and compare different support systems in terms of cost.
— Optimize the total cost of the excavation, or, in a mine, the production costs.

The total cost of any rockbolt reinforcement system can be divided into direct and indirect costs as shown below.

Direct costs

Direct costs are all costs included in the rock bolting activity until bolt installation is completed. The direct costs include the following components;

Material costs — bolt, face plate, grouting agent costs.

Material losses — some of the rock bolting materials sent underground will not be installed due to loss or damage. Lack of ruggedness and system complexity will lead to higher costs.

Material transportation — rock bolting materials are normally sold from a factory or distribution centre. Transportation costs to the work site will therefore be added. Distance, as well as the weight of the materials must therefore be considered in the cost evaluation of different systems.

Borehole costs — a number of factors, for example borehole diameter and depth, rock drillability, and type of drilling equipment used will affect drilling costs.

Installation costs — include the cost of installation equipment, for example dollies, drivers, maintenance costs on rock drills (if this is used to install the bolts), impact wrenches, torque wrenches and grouting or water pumps.

Labour costs — must be viewed on bolts per shift basis, as a considerable portion of the shift time can be spent in non-bolting related activities such as travel to work place, lunch rest etc.

Atlas Copco Roctec Inc. has developed a computer program for the evaluation of one or more bolting alternatives from a productivity and installed cost standpoint, given certain necessary input information related to type of drilling and installation equipment, rock type etc. Once the initial data has been loaded into the computer, the operator is free to change individual numbers and see the results of a change immediately.

For example, if it is possible to increase the drill penetration rate by 15%, what effect would this have on the total number of bolts installed per shift?

The program enables the responsible engineer to analyse each segment of the bolting cycle and pinpoint those that can be improved.

Indirect costs

Indirect costs are not related to the rock bolting as such. They depend on whether or not the reinforcement activity is an integral part of the excavation cycle. If the rock bolting is not an integral part of the excavation cycle, but is performed as a "stand alone" activity, the indirect costs of the rock bolting are the same as the "standstill costs" for the worksite.

Other examples of indirect costs are those which can be expected if rock bolting is not carried out, but a major rock fall occurs. These costs are related to the necessary extra mucking and reinforcement activities, production losses, as well as possible injury/damage to men and/or equipment. The cost of rock bolting must be weighed against the probability of a major rock fall as well as possible damage.

Indirect costs for rock reinforcement can be significant because of specific worksite conditions e.g. cleaning up after a major rock fall, combined with ore production losses at a critical time.

Minimize cost of rock bolting

The optimum reinforcement, from an economic point of view, is obtained when the sum of the direct and indirect costs are minimized. In the case of a major rock fall, the indirect costs may well be related to the direct costs, if the cause of the rock fall is insufficient rock bolting. Indirect costs of a rock fall cannot be quantified easily before the rock fall has occurred. However, this cost can be of the order of several magnitudes greater than adequate bolting costs. This implies that the most economic reinforcement is not governed by direct costs, but rather by the indirect costs. In severe rock mass conditions or a risk of high indirect costs, the direct costs become less important from an economic point of view.

A capacity and cost comparison between different types of equipment for rock bolting is presented in Part II of this handbook.

Chapter 10

References

D'ANDREA, FISCHER and FOGELSON, 1965.
Prediction of compressive strength from other rock properties. USBM RI 6702.

BARTON, N., 1973.
Review of a new shear strength criterion for rock joints. Engineering Geology, Elsevier, Vol. 7, 1973.

BARTON, N., LIEN, R. and LUNDE, J., 1974.
Engineering classification of rock masses for the design of tunnel support. Rock Mechanics, Vol. 6, no. 4.

BARTON, N., 1976.
Recent experiences with the *Q*-system of tunnel support design. Proc. Symposium on Exploration for Rock Engineering, Vol. 1, Johannesburg.

BARTON, N., LIEN, R. and LUNDE, J., 1977.
Estimation of support requirements for underground excavations. Proceedings 16th Symposium on Rock Mechanics, Minneapolis, USA.

BARTON, N., LØSET, F., LIEN, R. and LUNDE, J., 1980.
Application of *Q*-system in design decisions concerning dimensions and appropriate support for underground installations. Proceedings Int. Symposium (Rockstore '80), Vol. 2, Stockholm.

BEER, G., COWLING, R. and BYWATER, S., 1985.
A novel three-dimensional joint element with application to the modelling of mining excavations. Proceedings Int. Symposium on Fundamentals of Rock Joints, Björkliden, Sweden.

BERGMAN, S.G.A., KRAULAND, N., MARTNA, J. and Paganus, T., 1983.
Non-destructive field test of cement-grouted bolts with the Boltometer. Proceedings 5th Congress ISRM, Melbourne.

BIENIAWSKI, Z.T., 1973.
Engineering classifications of jointed rock masses. Trans S. Afr. Instn. Civ. Engrs. Vol. 15, no. 12.

BIENIAWSKI, Z.T., 1979.
The Geomechanics Classification in rock engineering applications. Proceedings 4th Congress ISRM, Montreux.

Bjurström, S., 1973.
Bergbultförband i sprucket berg. Fortifikationsförvaltningen, Forskningsbyrån, Rapport no. 121:3, Stockholm.
Bjurström, S., 1975.
Bergbultning — dimensionering, praxis och tillämpningar. BeFo Rapport nr 8, Stockholm.
Brady, B.H.G. and Brown, E.T., 1985.
Rock mechanics for underground mining. George Allan & Unwin, London.
Broch, E., 1983.
Estimation of strength anisotropy using the point-load test. Int. J. Rock Mechanics and Mining Sciences, Vol. 20, no 4.

Coulthard, M.A., Crotty, J.M. and Fabjanczyk, M.W., 1983.
Comparison of field measurements and numerical analyses of a major fault in a mine pillar. Proceedings 5th Congress ISRM, Melbourne.

Deere, D.U., 1964.
Technical description of rock cores for engineering purposes. Rock Mech. and Engng. Geol., Vol. 1, no 1.
Douglas, T.H. and Arthur, L.J., 1983.
A guide to the use of rock reinforcement in underground excavations. CIRIA Report 101, London.

Haas, C.J., Keith, H.D. and Dave, J., 1979.
Effects of expansion-anchor rock bolts on the stability of layered mine roofs. Mining Congress Journal, Sept. 1979.
Hanna, T.H., 1982.
Foundations in tension, ground anchors. Trans Tech Publications, Clausthal, Germany.
Hoek, E. and Brown, E.T., 1980.
Underground excavations in rock. The Institution of Mining and Metallurgy, London.
Hoek, E., 1982.
Geotechnical considerations in tunnel design and contract preparation. Trans Inst. Min. Metall. 91, A101—9.
Holmberg, R., Olsson, M., Sjöberg, C. and Smedberg, U., 1983.
Undermarksbyggande i svagt berg. 2A Borrning och sprängning. BeFo 1983.

Lama, R.D. and Vutukuri, V.S., 1978.
Handbook on mechanical properties of rocks, Vol. II. Trans Tech Publications, Clausthal, Germany.
Laubscher, D.H. and Taylor, H.W., 1976.
The importance of geomechanics classification of jointed rock masses in mining operations. Proc. Symposium on Exploration for Rock Engineering, Vol. 1, Johannesburg.
Ludvig, B., 1984.
Shear tests on rock bolts. Proc. of the Int. Symposium on Rock Bolting, Abisko, Sweden.

MILLER, H.D.S., 1984.
An evaluation of the Swellex rockbolting system. Atlas Copco Reg. code AHB TAB 57, Stockholm.
MORLAND, R.L. and THOMPSON, J., 1985.
Developments in ground support practices at Isa Mine. Proc. The Aus. I.M.M. Kalgoorlie Branch, Underground Operators Conference, October 1985, Australia.
ORTLEPP, W.D. and GAY, N.C. 1984.
Performance of an experimental tunnel subjected to stresses ranging from 50 MPa to 230 MPa. Proc. Symposium on Design and Performance of Underground Excavations, Cambridge, U.K.
PANDE, G.N., 1985.
A constitutive model of rock joints. Proceedings Int. Symposium on Fundamentals of Rock Joints, Björkliden, Sweden.
PANEK, L.A., 1964.
Design for bolting stratified roof. Transactions, Soc. of Mining Engineers, June 1964.
PENG, SYD S., 1978.
Coal Mine Ground Control. John Wiley & Sons, Inc. New York.

RABCEWICZ, L. and GOLSER, J., 1973.
Principles of dimensioning the supporting system for the "New Austrian Tunnelling Method", Water Power, March 1973.
RISSLER, P., 1978.
Determination of water permeability of jointed rock. Publ. Inst. Found. Engng., Soil Mech., Rock Mech. and Water Ways Constr., Vol. 5, RWTH (University) Aachen, Germany.

SCHACH, R., 1971.
Nye erfaringer med fjellbolter. Konferense i Fjellsprengningsteknikk, Oslo.
SCOTT, J.J., 1984.
Friction Rock Stabilizer impact upon anchor design and ground control practices. Proc. of the Int. Symposium on Rock Bolting, Abisko, Sweden.
SINGH, R.N., 1985.
Private communication. University of Nottingham. Dept. of Mining Engineering, Nottingham, England.
STILLBORG, B. and LEIJON, B., 1982.
A comparative study of rock stress measurement at Luossavaara mine. FB 8217 Swedish Mining Research Foundation.
STILLBORG, B., 1983.
Rock mechanics applications to open stope design at the Luossavaara research mine in Kiruna, Sweden, Status Report no. 1 FB 8325 Swedish Mining Research Foundation.
STILLBORG, B., 1984.
Experimental investigation of steel cables for rock reinforcement in hard rock. Doctoral Thesis. Division of Rock Mechanics, Luleå University of Technology, Luleå, Sweden.

STILLBORG, B., 1985.
Rock mass response to large blast hole open stoping. Proc. Symposium on Large Scale Underground Mining, Luleå, Sweden.
STILLBORG, B., 1986.
Utvärdering av Flexirope för kabelbultning. Internal report, JAA AB, Luleå, Sweden.

U.S. Army Corps of Engineers, 1978.
Shafts and Tunnels in Rock. Engineer Manual EM 1110-2-2901, 1978. Available from the Office of the Chief of Engineers, Washington DC, 20314.
U.S. Army Corps of Engineers, 1980.
Engineering and design: rock reinforcement. Engineer Manual EM 1110-1-2907, 1980. Available from the Office of the Chief of Engineers, Washington DC, 20314.

WAHLSTRÖM, E.E., 1973.
Tunnelling in rock. Elsevier Scientific Publ. Comp. Amsterdam.

Appendix

Input Data Form:
Geomechanics Classification of Jointed Rock Masses (Revised)
After Bieniawski (1979)

Input Data Form:
Geomechanics Classification of Jointed Rock Masses (Revised)
After BIENIAWSKI (1979)

Name of project:
Site of survey:
Conducted by:
Date:

STRUCTURAL REGION	ROCK TYPE AND ORIGIN

DRILL CORE QUALITY R.Q.D.*	WALL ROCK OF DISCONTINUITIES
Excellent quality: 90—100 %	Unweathered ....
Good quality: 75—90 %	Slightly weathered ....
Fair quality: 50—75 %	Moderately weathered ....
Poor quality: 25—50 %	Highly weathered ....
Very poor quality: <25 %	Completely weathered ....
*R.Q.D. = Rock Quality Designation	Residual soil ....

GROUND WATER	STRENGTH OF INTACT ROCK MATERIAL
INFLOW per 10 m of tunnel length litres/min	Designation — Uniaxial compressive strength, MPa OR Point-load strength index, MPa
or	Very high: Over 250 ..... >10 .....
WATER PRESSURE kPa	High: 100—250 ..... 4—10 .....
or	Medium high: 50—100 ..... 2— 4 .....
GENERAL CONDITIONS (completely dry, damp, wet, dripping or flowing under low / medium or high pressure):	Moderate: 25— 50 ..... 1— 2 .....
....	Low: 5— 25 ..... <1 .....
	Very low: 1— 5 .....

SPACING OF DISCONTINUITIES

		Set 1	Set 2	Set 3	Set 4
Very wide:	Over 2 m	....	....	....	....
Wide:	0,6—2 m	....	....	....	....
Moderate:	200—600 mm	....	....	....	....
Close:	60—200 mm	....	....	....	....
Very close:	<60 mm	....	....	....	....

STRIKE AND DIP ORIENTATIONS

Set 1	Strike: .... (average)	(from ....to ....)	Dip: .... (angle)	.... (direction)
Set 2	Strike:	(from ....to ....)	Dip:	....
Set 3	Strike:	(from ....to ....)	Dip:	....
Set 4	Strike:	(from ....to ....)	Dip:	....

NOTE: Refer all directions to magnetic north.

Input Data Form:
Geomechanics Classification of Jointed Rock Masses (Revised)
After BIENIAWSKI (1979)

CONDITION OF DISCONTINUITIES

PERSISTENCE (CONTINUITY)		Set 1	Set 2	Set 3	Set 4
Very low:	<1 m	....	....	....	....
Low:	1—3 m	....	....	....	....
Medium:	3—10 m	....	....	....	....
High:	10—20 m	....	....	....	....
Very high:	>20 m	....	....	....	....

SEPARATION (APERTURE)

Very tight joints:	<0,1 mm	....	....	....	....
Tight joints:	0,1—0,5 mm	....	....	....	....
Moderately open joints:	0,5—2,5 mm	....	....	....	....
Open joints:	2,5—10 mm	....	....	....	....
Very wide aperture:	>10 mm	....	....	....	....

ROUGHNESS (state also if surfaces are stepped, undulating or planar)

Very rough surfaces:	....	....	....	....
Rough surfaces:	....	....	....	....
Slightly rough surfaces:	....	....	....	....
Smooth surfaces:	....	....	....	....
Slickensided surfaces:	....	....	....	....

FILLING (GOUGE)

Type:	....	....	....	....
Thickness:	....	....	....	....
Uniaxial compressive strength, MPa	....	....	....	....
Seepage:	....	....	....	....

MAJOR FAULTS OR FOLDS

Describe major faults and folds specifying their locality, nature and orientations.

GENERAL REMARKS AND ADDITIONAL DATA

NOTE:
(1) For definitions and methods consult ISRM document: *'Quantitative description of discontinuities in rock masses.'*
(2) The data on this form constitute the minimum required for engineering design.
The geologist should, however, supply any further information which he considers relevant.

PART II*

Chapter 11

Atlas Copco Auxiliary Equipment for Rock Bolting

The Atlas Copco group of companies is one of the world's leading producers of equipment for compressed air, mining and construction equipment.

Atlas Copco is represented by subsidiary companies in more than 50 countries and by authorized dealers in another 85 countries. Each subsidiary company has its own local sales and service organization.

Further information on Atlas Copco products and applications may be obtained through your local representative.

The information presented in Part II of this handbook is a description of selected Atlas Copco equipment and its applications in rock reinforcement. A general description of the Swellex system is presented along with some case studies from mining and civil engineering. Part II of this handbook also deals with mechanized scaling, drilling and bolting. Basic principles, applications and economic considerations are covered.

Atlas Copco has implemented an intensive technical development program in the area of rock reinforcement. The direction these developments will take in the future is considered in the final section.

---

* by Mr. Robert Robertson, Mr. Björn Kåreby, Mr. Lorne Herron, Mr. Ken Lundgren and Mr. Anders Sylvan all at Atlas Copco MCT AB.

11.1 Swellex in Mining and Civil Engineering

11.1.1 General Description of the Swellex System

The Swellex system was developed in the Atlas Copco Research Department between 1977 and 1980. Extensive field tests were conducted, and in 1982 the Swellex system was put on the market. In 1986, the Swellex system will be available to all major mining and underground construction markets, with production plants located in Sweden, Canada and Australia.

The Standard Swellex system for rock reinforcement consists of a steel tubular bolt, an installation rod and a pneumatic high pressure water pump. Details of the system are given in Fig. 34.

The 2 mm thick, 41 mm diameter steel tube is mechanically folded to an external diameter of 26 mm. A cross section of the tube is shown in Fig. 34. Bushings are pressed on to the ends, which are then sealed by welding. The lower bushing has a flange to hold the face plate in place. Water under high pressure is injected into the bolt through an injection hole in the lower bushing. This causes the steel tube to expand and to adapt its shape to the irregularities of the borehole. As the swelling proceeds, the lower part of the steel tube shortens, pulling the face plate firmly (approx. 1.5 tons) against the rock surface. The installation pump automatically stops at the preset inflation pressure. The water pressure is released after installation, and the water drains out of the expanded steel tube.

A standard inflation pressure of 30 MPa (300 bar) gives the Swellex a bond strength to the rock of 10—20 tons per metre depending on rock conditions. The ultimate tensile strength of the bolt is 13 tons. The inflation pressure can be lowered to reduce the bond strength and allow the Swellex to slide, so that large rock movements may be accommodated. The Swellex can therefore be adapted to a variety of ground conditions by changing the inflation pressure.

A recent development of the Swellex system includes a yieldable Swellex, with an ultimate axial strain of 16%, compared to the 10% of a Standard Swellex. A corrosion protected Swellex, resistant to even very acidic ground water for longer periods of time, was also released in 1986.

The principal features of the Swellex rock reinforcement system may be summarised as follows:

- Bolt anchorage is relatively insensitive to variations in the borehole diameter (32—38 mm). It is, however, best suited for a 38 mm diameter borehole, a very common borehole size, which can be drilled using either light or heavy rock drills.
- Correct installation of the bolt does not rely on operator judgement and skill.
- Very simple installation. Installation quality can be virtually guaranteed.
- One man easily sets 40 to 50 bolts per hour in pre-drilled holes.
- Consistently high installation rates, give low unit costs.

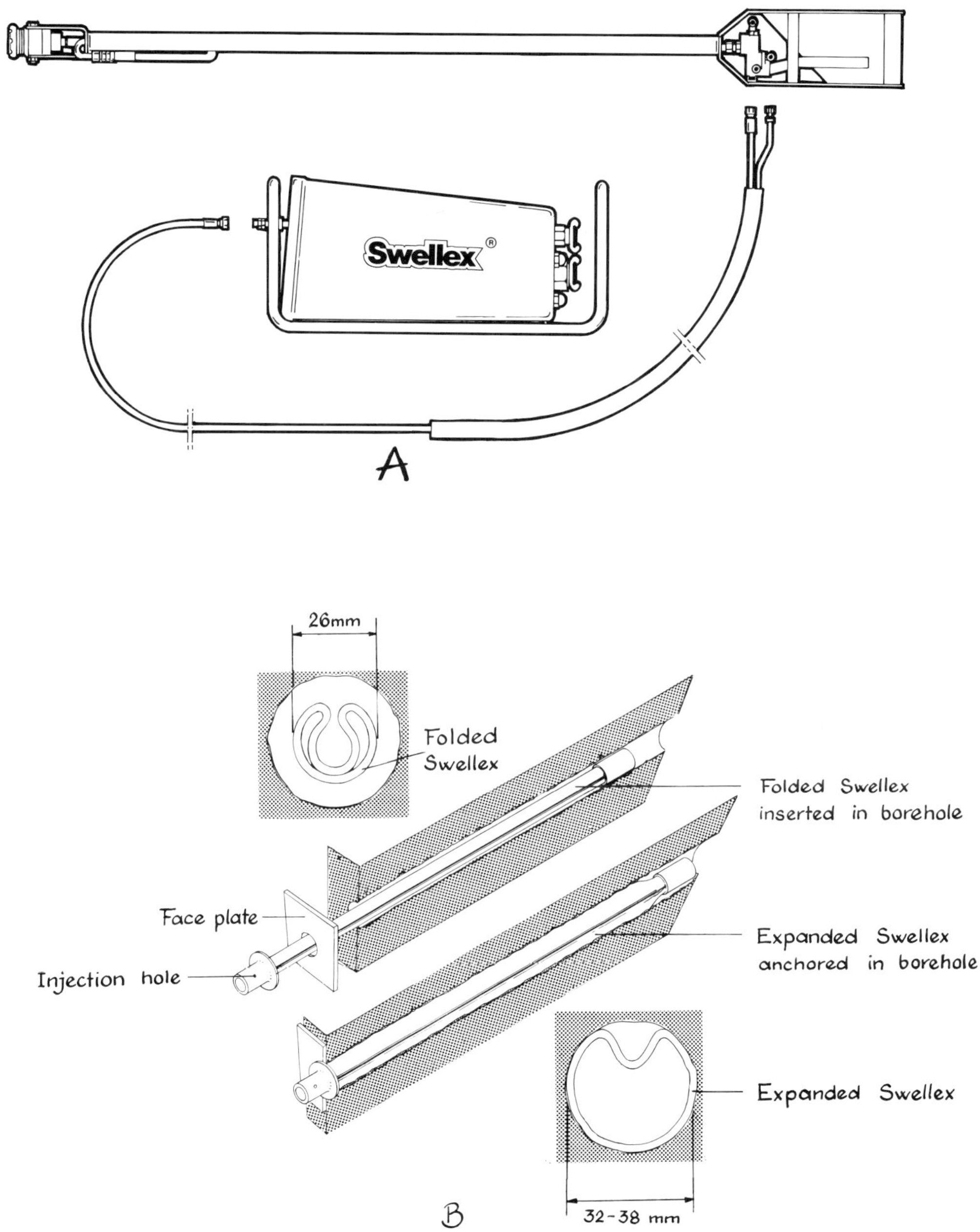

Fig. 34 Details of Swellex rock reinforcement system.
A. Pneumatic high pressure water pump with installation arm.
B. Installation
Partly after HOEK (1982).

— Any required length of Swellex is available, and the installation procedure is not affected by bolt length.
— The bolt is equally suited for manual and mechanized bolt installation.
— Immediate and full load bearing capacity can be virtually guaranteed for each bolt, since the pump automatically stops at the preset inflation pressure.
— The bolt can be adapted to a variety of ground conditions, see also section 6.4 for further details.
— Unlike grouted bolts, the performance of Swellex is not adversely affected by joints in the rock mass, but the bond strength increases when the bolt deforms into these joints.
— Some 200 mines and construction sites are currently (1986) using the Swellex system.

11.1.2 Swellex — A Versatile System for Rock Reinforcement

Pit wall and slope stabilization

Pit wall and slope stabilization present special problems in ground control. Blasting, water accumulation and slippage cause the opening of cracks whose width sometimes exceeds one foot. These cracks generally run parallel to the pit wall making the drilling of rockbolt holes difficult. Successful installation of fully grouted anchors is unsafe and nearly impossible under these conditions. The installation of wiremesh to contain local loose rock only compounds this difficult operation.

Swellex rockbolts in lengths of up to 9 m (30 ft.) are used in Canadian open pit mines for pit wall stabilization. The Swellex bolt is slid manually into a pre-drilled hole and expanded with a high pressure water pump to provide anchorage of up to 20 tons per metre. It has the advantage of bridging open cracks and anchors the rock on either side of the crack. It reforms and contacts the rock along its full length up to the collar of the hole, providing full support on the pit wall face.

This technique minimizes costly scaling and cleaning up which is normally required before bolting can take place. Wiremesh can also be easily installed at any time on pre-installed bolts with the simple mesh washer available with Swellex. The mesh is simply suspended and washers are installed without drilling additional holes.

Since the bolt is tubular in shape, the Swellex tolerates up to 35 mm (1 3/8″) of shear movement before tensile loading of the bolt is apparent. A 7 m (24 ft.) long Swellex bolt takes approximately 3 minutes to install in a pre-drilled hole. Cost and productivity savings of up to $50 per bolt on grouted rockbolt installation are obtained. Figure 35 illustrates some of the basic features (described above), of the Swellex system for rock stabilization.

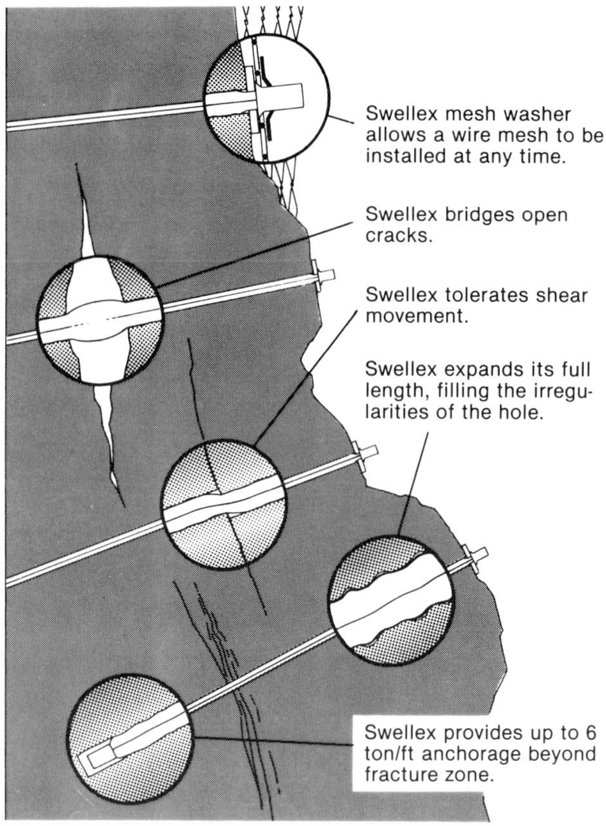

Fig. 35 Illustration of some basic features of the Swellex system for rock mass stabilization of pit walls and slopes.

Long Swellex bolts

Long Swellex bolts have also found applications in underground mining. With new mining methods, the demand for rock stabilization with very long bolts is increasing. Although cable bolts and extension rebars are normally used in such applications, many users find greater advantages in longer-than-drift-height Swellex rockbolts.

Stabilization of brows in sublevel caving and backs in cut and fill mining are typical applications, where up to 9 m (30 ft.) long Swellex rockbolts can be installed in headrooms as low as 2.7 m (9 ft.). Swellex is ideal for such applications for the following reasons:

— Large torque or feed force are not required for Swellex installation.
— Swellex rockbolts are sufficiently flexible to be installed in headrooms with a much smaller roof height than bolt length.

— Swellex is expanded by water under high pressure, forcing it to grip the rock along its entire length. 100% effective bolt installation is obtained even in heavily fractured ground.
— Swellex bolt installation gives exceptionally good resistance against blast generated vibration. It can therefore be used close to the advancing rock face.

Swellex long rockbolt installation can be performed more rapidly and far cheaper than the installation of other long rockbolt or cable bolt systems. In a recent underground installation, the drilling time for a 7.5 m (25 ft.), 38 mm (1½″) hole was 8 min. Installation of a 7 m (24 ft.) long Swellex bolt only required 3 min. installation time for a 2 man crew. The installation procedure for a longer-than-drift-height Swellex rockbolt is illustrated in Fig. 36.

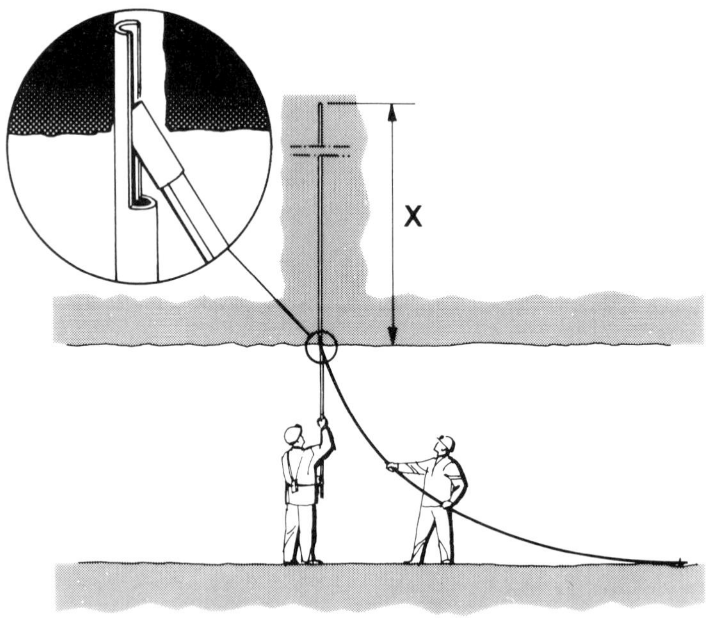

Fig. 36 Installation procedure for a longer-than-drift-height Swellex rockbolt.

Sliding Swellex

By adjusting the standard inflation pressure of 30 MPa (300 bar), the bolt can be adapted to a wide variety of ground conditions. In ground where large rock movements can be expected, the inflation pressure can be reduced by some 20—30%. This reduces the bond strength, the bolt can slide and hence accommodate necessary deformation of the rock. At any level of rock deformation, the bolt can be repressurized to the standard inflation pressure of 30 MPa (300 bar) to provide full bond strength.

Wire rope lacing with Swellex

Situations can arise in underground mining where fairly large masses of fractured rock cannot be retained by rockbolts alone. In these cases the normal alternatives are to either remove the unstable rock or to support it externally. Swellex rockbolts now offer an alternative solution to this problem.

By drilling holes on either side of the loose rock, wire ropes can be stretched across it and anchored by Swellex bolts, see Fig. 37. 45 mm (1 $^3/_4$″) diameter holes are drilled and 12 mm ($^1/_2$″) diameter wire ropes are used. One end of the wire rope is inserted into a hole alongside a Swellex bolt, which is expanded to anchor the wire rope. The other end of the rope is bent and threaded into a hole on the other side of the loose formation. A T-bar is placed in the fold of the rope and the Swellex bolt is used to push it into the hole, (see also Fig. 37). The force from the rock drill feed can be applied against the lower bushing to tension the wire rope before the Swellex bolt is expanded.

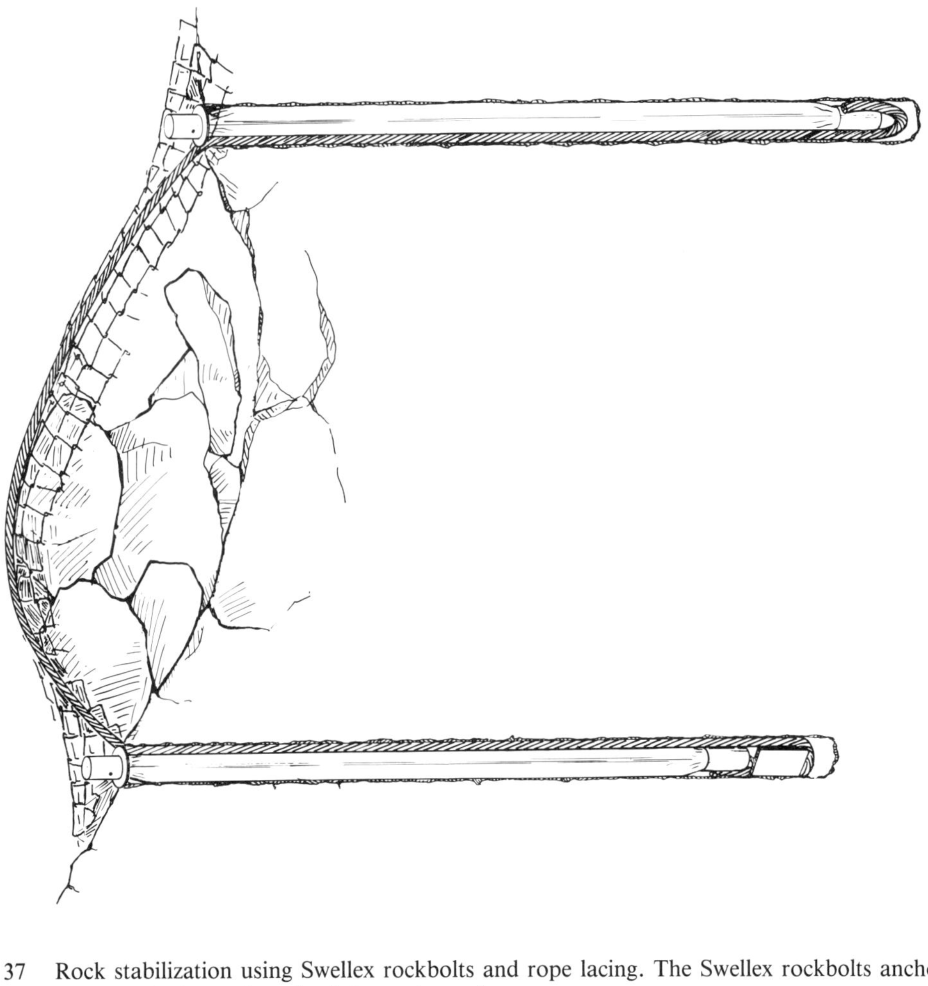

Fig. 37 Rock stabilization using Swellex rockbolts and rope lacing. The Swellex rockbolts anchor the rope in the rock and reinforce the rock.

The same procedure can be used to install cable trusses in coal mine entries, see Fig. 38.

Rope lacing may also be used to stiffen wire mesh in cases where the mesh cannot constrain the loosened rock adequately.

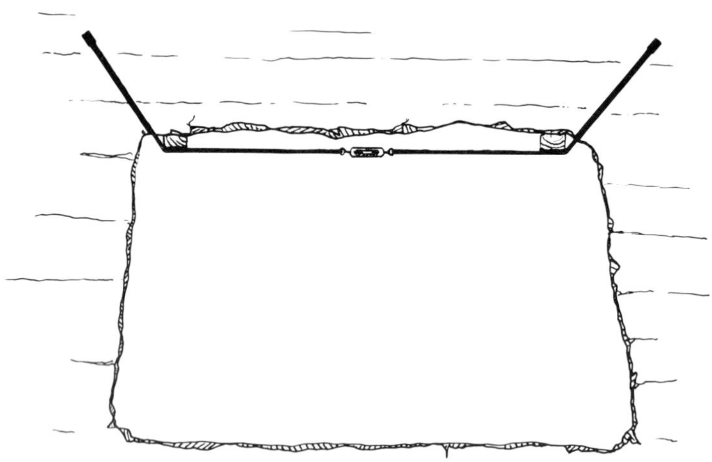

Fig. 38 Rock stabilization using cable truss. The Swellex rockbolts are used to anchor the cable in the rock.

Swellex in rock grouting

In tunnelling under very difficult ground conditions, and when the rock carries a lot of water, pre-grouting of the rock ahead of the advancing tunnel face is sometimes required. With water flowing from drill holes under high pressure, rock grouting can be virtually impossible. Under these conditions, conventional hole plugging with sleeves is found to be inadequate. In an alternative method, Swellex rockbolts are used as a plug to seal the hole and the grout is then injected through the bolt. In this method the sleeve consists of a Swellex rockbolt covered by a PVC hose, and a hydraulic tube inserted through the bolt. A seal is provided when the bolt is inserted into the collar of the hole and expanded. The grout can then be injected through the hydraulic tube which now runs alongside and past the Swellex rockbolt. Details of a Swellex seal for rock grouting are shown in Fig. 39.

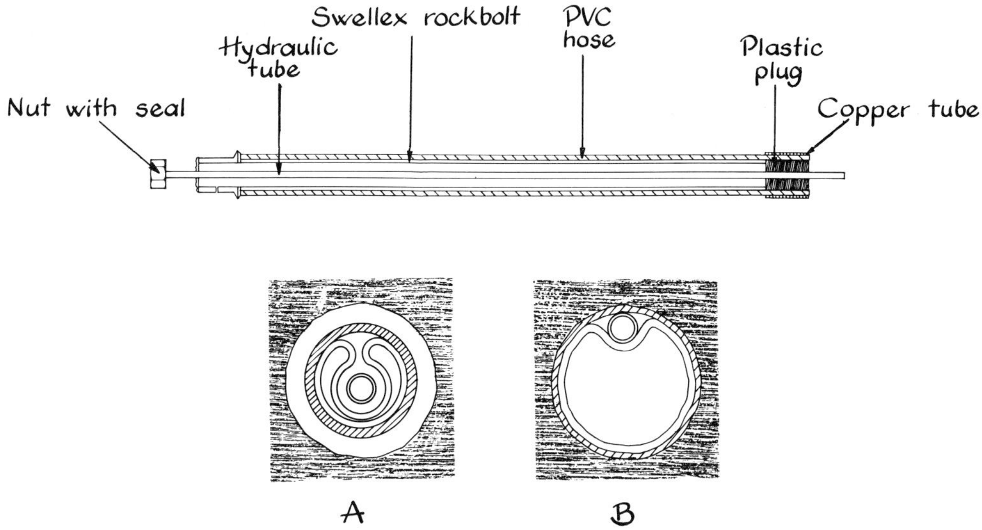

Fig. 39 A Swellex rockbolt converted to a grouting collar for rock grouting.
A. Folded Swellex with hydraulic tube inside and surrounded by a PVC hose.
B. Expanded Swellex with hydraulic tube on its periphery, borehole sealed.

11.1.3 Case Histories of Swellex in Mining and Civil Engineering Applications

Swellex in sericitic schist

A Canadian gold mine faced with very difficult ground conditions where the orebody and immediate country rocks are sericitic and highly schistose. The intensity of schistosity varies significantly, and poor ground conditions prevail, especially within the orebody.

The ore has been mined by using a modified sub-level caving technique as the principal mining method. The ore is broken by rings drilled upwards from sub-levels 9—15 m (30—50 ft.) apart.

Failures in the drifts assume the form of buckled schistosity planes. This type of failure is generally characterized by the accumulation of small deformations, until a stress "trigger level" is reached, followed by rapid, large scale movements as the slabs buckle. This sometimes occurs within two weeks of development. The key to good drive support is to keep the sidewall intact.

In 1983 the mine switched from resin grouted rebars to Swellex bolts which were considered easier to install correctly and have a higher degree of worker acceptance. Much of the rehabilitation work previously conducted in the mine has been eliminated. The mine has also become the first underground mine to undertake extensive supporting by using longer-than-drift-height Swellex bolts. These bolts are being used to support areas previously requiring steel arches. According to reports from the mine engineers, Swellex has substantially reduced ground support costs.

Swellex in coal mining

Swellex rockbolts are used in British coal mines where wooden dowels or resin grouted rockbolts were previously used. In particular, the miners have found the Swellex rockbolt useful in the support of salvage areas in retreat mining applications, where the bolts are applied in combination with weld mesh.

In comparison with the traditional methods for supporting the salvage area, the Swellex bolt and the weld mesh offers the following advantages:

— Safer working conditions.
— Preparation time of the salvage area at the face end is halved.
— Material costs halved.

Swellex has also found application as a general support tool in a Norwegian coal mine at Spitzbergen. Here, in the room and pillar mining operation, Swellex is used to support the roof in conditions where the preheated ventilation air softens the permafrozen ground. The expanded shell anchored rockbolts used previously, were replaced since the support provided by the Swellex bolt was superior, and the installation of the 1.8 m Swellex bolt could easily be mechanized.

Swellex rock bolting from TBMs

Rock bolting from tunnel boring machines presents special problems. The rapid advance of these machines, and the confined space in which the work must be performed, require a well-developed, integrated system.

To date there are several projects in which Swellex rock bolting is combined with TBMs. In two major projects in Canada e.g. Swellex rockbolts are being installed from TBMs — the Montreal sewage tunnels and the Rogers Pass railway tunnel.

The Montreal sewage tunnels are driven in predominantly limestone rock containing clay seams and some shale. The line of the tunnel intersects the limestone layers at a shallow angle.

The tunnel is bored to a diameter of 6.15 m (20 ft. 2 inches) in a single pass, using a Robbins 203/205 TBM by the Montreal based contractor, Fitzpatrick Construction Ltd. Feed mounted hydraulic rock drills are pivotly mounted to the outer frame of the machine just forward of the grippers, allowing the holes to be drilled while the tunnel is bored. The Swellex bolts are then installed manually. The 2.1 m (7 ft.) bolts are installed on a 1.8 m by 1.8 m (6 ft. by 6 ft.) pattern. Whenever strapping or wire mesh is required, the installation chuck can be mounted on a dolly which is inserted into the rock drill, allowing the feed force to be used to form the mesh or strapping to the contour of the tunnel.

The Rogers pass tunnel, Selkirk side, is being bored to a diameter of 6.8 m (22 ft. 4 in.). The bottom is then excavated to give a horseshoe profile. Bolting is conducted as required with stopers used to drill the holes. The 2.4 m (8 ft.) long Swellex bolts are installed manually.

According to the respective contractors from both of these work sites, the ease and speed of installation of Swellex bolts has reduced the number of delays (normally attributed to rock control) appreciably.

In a large French TBM tunnel project, Swellex is used to anchor wire mesh and steel sets to the rock at close intervals in poor ground conditions. The TBM has a high advancing rate (up to 6 metres an hour). The Swellex system provided the contractor with a reinforcement system whose installation was sufficiently rapid to keep pace with the TBM.

Tests at the Neuenburg highway tunnel project in Switzerland have shown that the cutter heads on a TBM or a roadheader easily cut through the Swellex rockbolt without damaging the cutters.

Swellex in soft rock applications

Rock bolting in soft rocks has always created stability and safety problems. A big Italian contractor faced very severe problems at a freshwater project in southern Italy. When one of the tunnels was driven through homogeneous limestone, he suddenly encountered a clay zone which made further drilling and blasting impossible.

The major problem here was to find bolts which could anchor the steel arches in the soft material. Tests were made with several alternative bolt types. Swellex proved capable of withstanding considerably higher loads than any other bolt type. It was therefore considered ideally suited for this task.

The average pull-out force of the 3.0 m testbolts was initially reported to be 8.5 tons which, however, increased to 9.6 tons some 5 days later, because water in the clay had drained through the bolthole.

The second problem involved stabilization of the tunnel face and the prevention of clay swelling into the tunnel. Again tests were made with Swellex. This resulted in a new (and unusual) method for tunnel driving in clay formations:

— Some 40 Swellex bolts of 3.0 m length were set in predrilled holes into the tunnel face. 1.5 m of the stabilized tunnel was then chipped out, using a scaling unit with a 0.5 m digging tooth. The bolts remained in the holes but were bent in all directions.
— The tunnel walls were then supported by steel arches anchored with Swellex.
— Another 40, 3.0 m Swellex bolts were installed just below the previous ones and a further 1.5 m of the clay was chipped out. According to the contractor, the Swellex rockbolt helped him arrive at a prompt, simple and economical solution to his unexpected problems.

Swellex in karst

Karst is a type of limestone partly eroded by water. This creates severe problems for conventional rock bolting.

Swiss contractors found that neither expansion anchored rockbolts nor grouted rockbolts could provide reliable ground support.

Since Swellex is unaffected by open cracks or voids, it was selected as the best choice.

Swellex in sandstone

Molasse is a soft sandstone with a compressive strength of 40—80 MPa (400—800 bars). In 1983 a Swiss contractor started a tunnel project under the town of Lausanne. A roadheader was used due to the soft molasse rock.

For the temporary reinforcement, Swellex rockbolts were used, together with wire mesh and shotcrete.

The installation of common rockbolt types is difficult since the borehole diameter varies considerably in the soft rock.

A test program was conducted and different bolt types were tested. The contractor decided to use Swellex. His decision was principally based on the following observations:

— An average pull-out resistance of 11 tons per metre of bolt length was obtained.
— Immediate support action. Simplicity of installation.

Swellex in a New York freshwater tunnel

A 6 km section of a fresh water tunnel was driven through a schisty granite-gneiss rock type in 1983—84 from upstate New York to lower Manhatten.

The consultants originally specified cement grouted rebar as part of the permanent reinforcement system. However, after extensive testing it was decided to change to Swellex because of its ease of installation and immediate support action. After the rock bolting, a 50 cm concrete lining was applied. The contractor reported significant time saving and cost reduction thanks to the revised method.

11.2 Mechanical Equipment for Scaling

Scaling and barring down are classic, traditional operations in mining and underground excavation where considerable emphasis is placed on the miners' skill, judgement, experience and "feel" for hanging loose rock. It also requires considerable physical effort. The trend to heavier mechanized drilling equipment and higher, wider faces and ends, pushed the possibility of hand scaling to its limits in many cases. It certainly made the work harder and more costly in terms of manhours expended in scaling.

Furthermore, the use of baskets on extendable booms so that scalers could reach high backs introduced its own additional safety hazards. It is hard to run away from a falling block of rock when you are standing in a basket several metres up in the air. Also serious accidents have been caused when falling rocks overturn the scaling platform. Mechanized scaling has developed over the last decade to make this safer and more efficient than hand scaling. Mechanized scaling rigs, often developed as a result of close collaboration between mining companies, contractors and equipment manufacturers, have resulted in fewer accidents to men, reduced damage to equipment, and higher output from the stope or development end. Studies conducted at the Boliden Mineral AB in mines using different mining techniques, have shown that compared with hand scaling, mechanized scaling gives better quality results, is much faster, is 25 to 40 percent more efficient, and is a superior method from the point of view of safety, production and human ergonomics.

There are three different principles of mechanized scaling, breaking, scratching and impact hammer. Breaking uses the same principle as that used in hand scaling. However, since very good visibility is a necessity for this method, it is not frequently used. The principle of scratching has the disadvantage that the reactive forces of the scaling action are transmitted back through the arm and the machine body. This leads to high maintenance costs and low machine availability due to frequent breakdowns. The scratching type scalers are only favoured for horizontally stratified roofs in weak rock formations. Both breaking and scratching scalers require stable and heavy booms and carriers.

Impact type scalers have a light hydraulic hammer which is mounted on a flexible boom and light carrier. Because the hammer is light, it can be mounted on a long boom, which gives the scaler a wide coverage area. The impact force exerted by the hammer should be adjustable, so the hammer power can be matched with the rock. The scaler should not break rock out of the solid. These are also the basic features of the Atlas Copco equipment for mechanized scaling, illustrated schematically in Fig. 40.

11.3 Mechanical Equipment for Rockbolt Installation

Considering the equipment used, rockbolting can be divided into three main groups;

A. **Manual equipment;** very light hand held drilling and installation equipment.

B. **Semi-mechanized equipment;** mechanized drilling and manual equipment for rockbolt installation.

C. **Fully mechanized equipment;** mechanized drilling and rockbolt installation controlled from an operator's panel placed under a safe roof.

A. Manual equipment:

In small drifts and tunnels, where drifting is performed with hand held equipment, the boltholes can be drilled using the same equipment. In larger area drifts and tunnels, stopers are commonly used for bolthole drilling, and the rockbolts are installed manually, for details see Fig. 41A. Service trucks, and trucks for blasthole charging are sometimes used both as a platform for hand held drilling of boltholes, as well as for rockbolt installation.

Some stoper models can also be used to spin the bolt into the cement or resin cartridges, and also to tension the nut at the bolt head. Special, so called nutrunners are also commonly used for this application. When the nut is to be tensioned to a pre-determined torque, a torque wrench is to be recommended. If the bolt is to be tensioned to a specified load, a hydraulic bolt tensioner with a calibrated gauge for load registration is preferred.

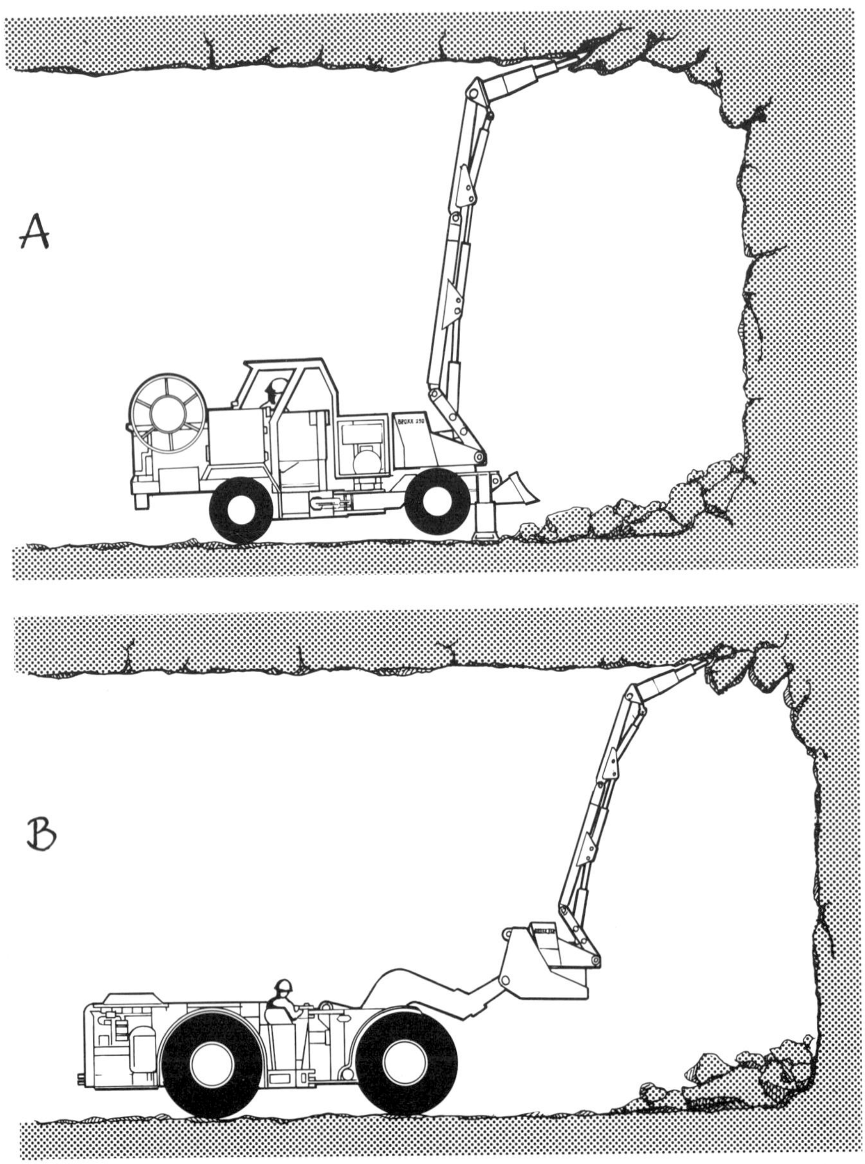

Fig. 40 Mechanical equipment for scaling.
A. Brokk PT 50 Scaler for roof hights between 3 and 6 metres.
B. Brokk scaling system mounted on an LHD for roof hights between 5 and 9 metres.

For coal mines, a special hand held rotary rock drill, a Pneumatic Roof Bolter (PRB), has been designed. The Roof Bolter is basically a rotary stoper which can also be used to install resin grouted bolts.

B. Semi-mechanized equipment:

In larger tunnels and in undergound excavations in general, where the feed height of a stoper is not sufficient, a drill jumbo can be used for bolthole drilling, see Fig. 41 B1. A big jumbo is sometimes also equipped with an extendable boom or working platform/basket, from which rockbolt installation can be conducted, see Fig. 41 B2. On a TBM or a roadheader, a hydraulic rock drill can be pivoted to the machine frame, to facilitate mechanical bolthole drilling, see Fig. 41 B3. Mechanized drilling with manual bolt setting is called "semi-mechanized rock bolting".

In confined areas, where the feed of the drill jumbo might be too long, but must nevertheless be used for the bolthole drilling, telescopic feeds are sometimes used. A telescopic feed enables the blastholes at the face to be drilled to normal length, in the extended position, and also the boltholes with the feed in collapsed position.

C. Fully mechanized equipment:

Fully mechanized rockbolting rigs drill the boltholes and install the rockbolts. This is controlled from an operator's platform. Two different fully mechanized rockbolting rigs are shown in Figs. 41 C1 and C2. The rig in Fig. 41 C1, is a fully mechanized bolting rig for Swellex bolts. The rig shown in Fig. 41 C2 has a system of one feed with a three position centralizer on a single boom. This is designed to install mechanically- grouted- and friction- anchored rockbolts. The first position is for the drilling of the bolthole, the second for injection of the cement grout or resin cartridges, and the third for the rockbolt installation. The system ensures that both the injection pipe for the grout and the bolt index accurately, and hit the bolthole. The principle of operation is illustrated schematically in Fig. 42.

Recent developments by Atlas Copco, in the area of mechanized equipment for rock bolting, include a fully mechanized cable bolting rig. The rig, illustrated in Fig. 43 A, has a number of features necessary for cable installation in any borehole direction, in lengths up to 50 metres. The rig shown in Fig. 43 A is equipped with an airconditioned cabin on a diesel driven carrier. Besides the electrohydraulic automatic drilling components, with rodhandling, it also includes the following cable bolting equipment;

— a hydraulically driven cable drum for 500 m of 28 mm diameter cable.
— a hydraulically driven cable feeder mounted on a hydraulic cylinder feed.
— a scissor table for 1 ton of cement.
— a hydraulically driven hosereel for 60 m cement grout hose/waterhose.
— a mixer for water-cement.
— a cement grout pump.

The cable bolting will therefore be a one-man operation.

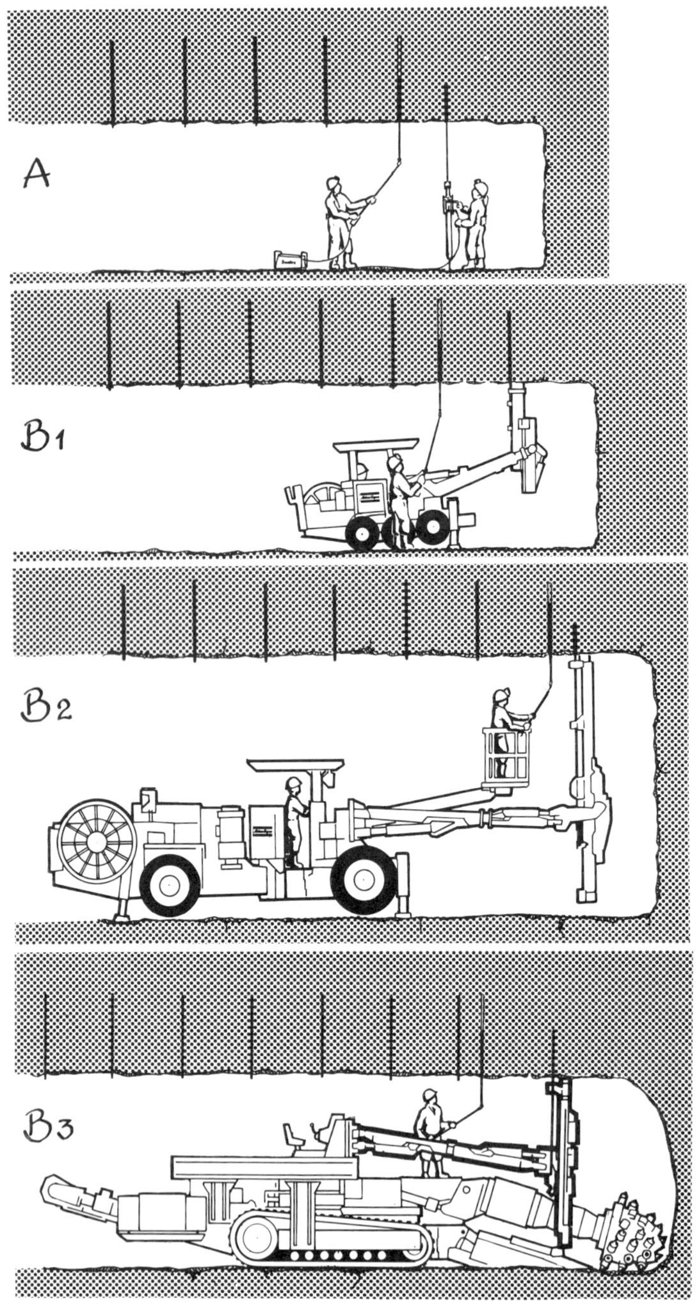

Fig. 41 A—B Equipment for rockbolting.

A. Manual rock bolting using a stoper rock drill and manual bolt setting.

B1. Semi-mechanized rock bolting using a small drill jumbo and manual bolt setting.

B2. Semi-mechanized rock bolting using a large drill jumbo with a basket for manual bolt setting.

B3. Semi-mechanized rock bolting using a boom and feed mounted hydraulic rock drill on a roadheader for manual bolt setting.

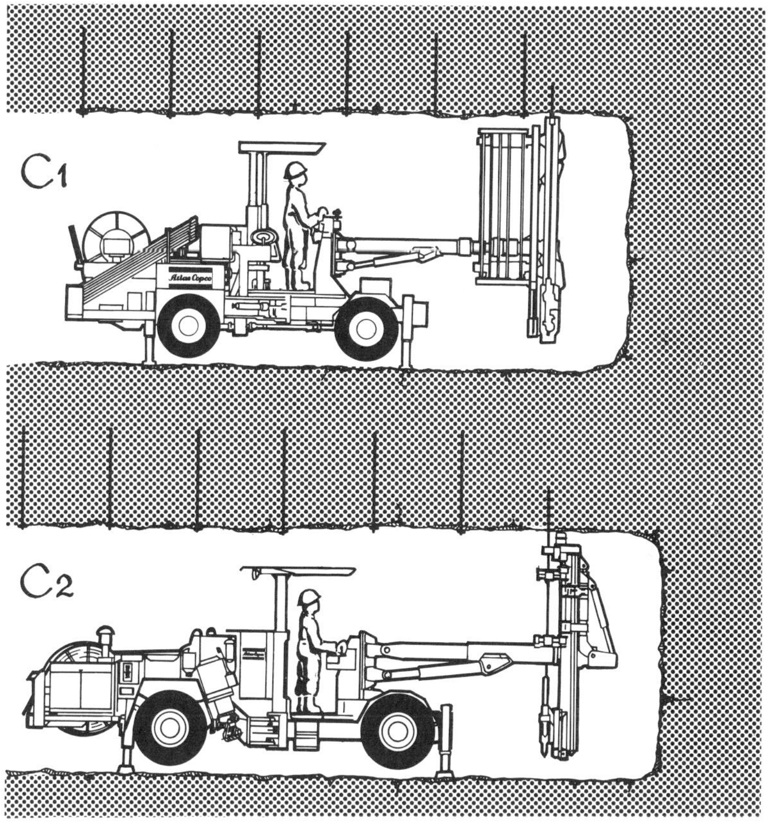

Fig. 41 C Equipment for rockbolting.
C1. Fully mechanized rock bolting using a rig for Swellex rockbolt installation — the Swellbolter.
C2. Fully mechanized rock bolting using a rig that can be equipped for any rockbolt type — the Boltec.

It is, however, sometimes convenient to have two separate units for the cable bolting operation, e.g. a standard hydraulic production drillrig for drilling the boltholes, and all equipment necessary for cable bolt installation placed on a separate unit, see Fig. 43 B. The advantage of this is that the drilling can be used independently of the cable installation unit in other drilling jobs, and the cable installation unit can, if necessary, support more than one drillrig.

Today (1986), it is estimated that some 50% of all rockbolting is performed with light hand held drilling equipment. Semi-mechanized bolting is estimated to represent about 40% of all rockbolting activities. Less than 10% of all rockbolting activity is fully mechanized.

Even though it is sometimes difficult to justify the expense of a fully mechanized bolting rig, more completely mechanized rockbolting will be performed in the future, for the following reasons;

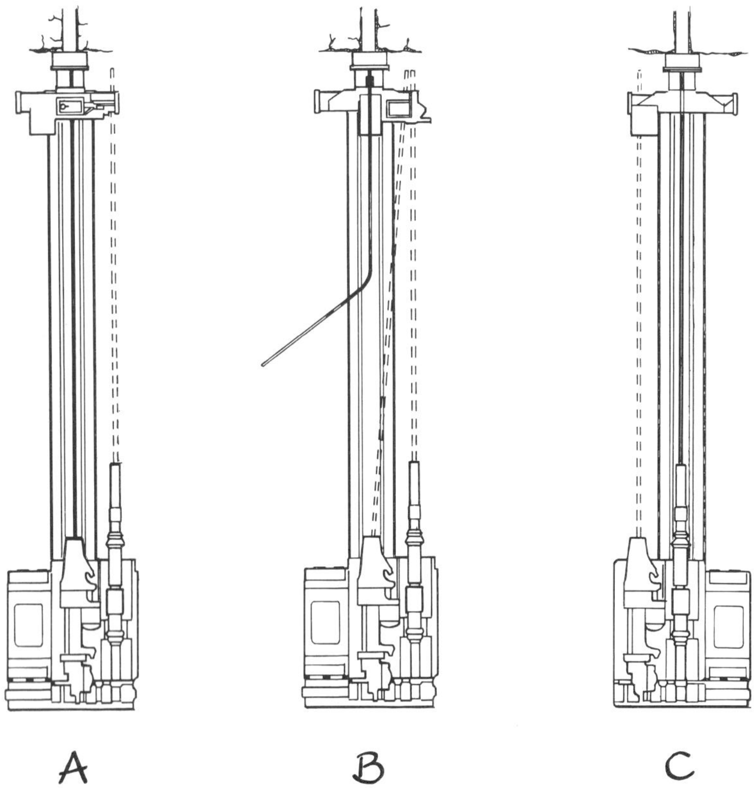

Fig. 42 Operation principle of the one feed system for grouted rockbolts. The three operation positions are as follows:
A. Bolthole drilling.
B. Injection of cement grout or resin cartridges.
C. Rockbolt installation.

1. Operator safety is significantly improved. During the entire operation, the operator is located at a safe distance from the bolting, under a roof that has already been bolted. He is therefore protected against the following:
   — falling rock
   — splash of grout, cement or resin in the eyes
   — hand held rock drill, pusher legs hazards
   — slippering
   — development of allergies from chemical spillage
   — fatigue.
2. Damage to equipment is minimized. Since each bolt is installed immediately after the hole has been drilled, and before the stinger of the bolting turret has been removed, (feeding pressure 1.5 ton) the risk of rock fall is minimized.
3. Installation quality of grouted rockbolts is improved.
   — with pneumatic charging of cement or resin cartridges, it is more likely that the bottom of the hole will be reached without cartridge damage ''half way up'' in the drilled hole.

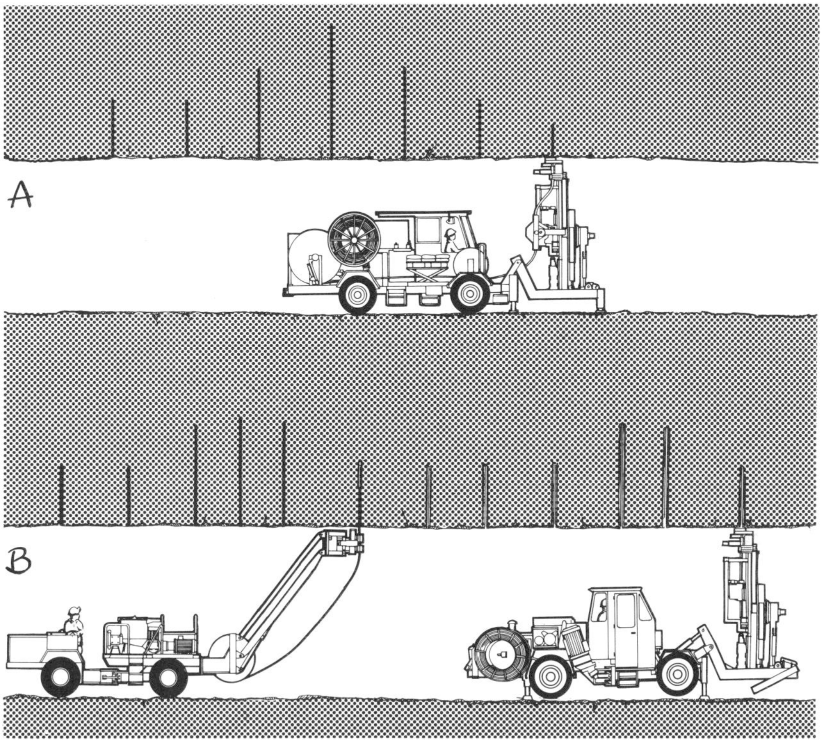

Fig. 43 Equipment for cable bolting in any borehole direction in lengths up to 50 metres.
A. Fully mechanized cable bolting rig.
B. A production drillrig combined with a cable installation unit.

— with the Atlas Copco's "Ceminject" system, bulk cement can be used. The high pressure injection of a thin (8 mm) string of cement will almost always reach the bottom of the bolthole despite irregularities in the hole.

4. Better reach. At present (1986), equipment for roof heights exceeding 12 m is available. For a practical roof height of 7.6—8.6 m, several bolts can be installed from the same machine set-up.

5. Increased bolt lengths. Bolting heads for bolts whose lengths range from 1.5 to 6.0 m are available.

6. Increased precision. Thanks to the modern design of the hydraulic system, booms and joy-stick controls, the positioning time between holes is minimized. The boom and bolting turret can be readily positioned and the bolt set parallel to other bolts if desired.

7. Increased speed. High speed is often required in ramps and on flat ground when one bolting unit is to serve several headings/faces. Versatile articulated and narrow carriers can now be selected for optimum underground mobility.

8. One man operation. Atlas Copco fully mechanized bolting rigs are now designed for one man operation. The operator is protected by a roof or an ergonomically designed cabin. (In some countries one man operation is prohibited by law. In such cases the second man (the helper), can carry out a number of activities. This increases the availability and productivity of the rig.)
9. Optimum long term capacity. The following factors, most of which are site specific, will affect the long term bolting capacity:

— systematic bolting or spot bolting	— equipment utilization
— bolt type	— work planning
— bolt length	— number of men in operation
— roof height	— maintenance routines
— rock type and rock mass conditions	— work continuity
— transport distance between worksites	— wages system
— operator experience and service organisation	— ambient temperature
— equipment reliability	— altitude.

11.3.1 Capacity and Cost Comparison Between Different Types of Equipment for Rock Bolting

An attempt is made to quantify the capacity and costs of different types of equipment for rockbolting. Two rockbolt types are considered: bulked cement grouted rockbolts and Swellex. The figures quoted in the capacity and cost calculations represent averaged field data. *The results presented should therefore be considered only as an index for comparison.* Depending on local conditions, rather different results may be obtained.

Equipment considered in capacity comparison: Hand held, Semi-Mechanized and Fully Mechanized.

Equipment considered in cost comparison: Semi-Mechanized and Fully Mechanized.

Capacity comparison

Pre-set Conditions

Rock type	: Granite
Hole length	: 2.5 m
Hole diam.	: 35 mm/38 mm
Bolt type	: Cement grouted rebar rockbolts and Swellex

Penetration rate:	Hole diam.		Average:
— Hand held BBD 90;	35 mm diam.	0.35—0.50 m/min	0.43 m/min
— Semi-Mech. COP 1032 HD;	38 mm diam.	0.85—1.40 m/min	1.10 m/min
— Fully Mech. COP 1032 HD;	38 mm diam.	0.85—1.40 m/min	1.10 m/min
COP 1028 HD;	38 mm diam.	0.65—1.15 m/min	0.90 m/min

Rock bolting operation time (in minutes) for different equipment, one man operation.

Operation	Hand held (1)	Semi-Mech. (2)	Fully Mech. (3)	Semi-Mech. (4)	Fully Mech. (5)
	Cement grouted rockbolts			Swellex	Swellex
Collering	0.25	0.15	0.15	0.15	0.15
Drilling	5.80	2.30	2.30	2.80	2.80
Return rock drill	0.40	0.20	0.20	0.20	0.20
Indexing for grouting	0.50	0.50	0.05	—	—
Hose feeding	0.40	0.40	—	—	—
Grouting	0.45	0.45	0.05	—	—
Indexing for bolting/handling	0.50	0.50	0.05	0.50	0.05
Bolting	0.50	0.50	0.35	0.50	0.35
Positioning for new hole	1.50	1.00	1.00	1.00	1.00
Net bolting time min/bolt	10.30	6.00	4.15	5.15	4.55
Net bolting capacity bolts/hour	5.8	10.0	14.5	11.7	13.2
Set-up time					
Moving rig/set up	—	0.10	0.10	0.10	0.10
Moving platform	0.20	0.20	—	0.20	—
Connect air, water electricity etc.	1.50	1.25	1.00	0.75	0.75
Walking/row	0.70	—	—	—	—
Change drillsteel	0.25	0.15	0.10	0.15	0.05
Preparations -mixing cement	0.45	0.45	0.40	—	—
-cleaning	0.90	0.90	0.60	—	—
Bolts in magazine	—	—	0.20	—	0.20
Service -daily/weekly	0.40	0.80	0.60	0.50	0.50
Stop caused by -the rig	0.30	0.75	0.50	0.45	0.45
-rock conditions	1.50	0.80	0.50	0.50	0.50

(Continued)

Set-up time	Hand held (1)	Semi-Mech. (2)	Fully Mech. (3)	Semi-Mech. (4)	Fully Mech. (5)
	Cement grouted rockbolts			Swellex	Swellex
Personnel time	0.70	0.50	0.40	0.70	0.30
Planning	0.10	0.10	0.10	0.10	0.10
Set-up time min/bolt	7.00	6.00	4.50	3.45	2.95
Total bolting time min/bolt	17.30	12.00	8.65	8.60	7.50
1 man operation					
Capacity bolts/hour	3.5	5.0	6.9	7.0	8.0
Shift capacity bolts/shift (6hrs.)*	21	30	41	42	48
Capacity relations in %	50%	71%	98%	100%	114%
2 man operation**					
Capacity bolts/hour	5.8	7.4	8.5	11.8	8.8
Shift capacity bolts/shift (6hrs.)*	35	44	51	71	53
Capacity relations in %	49%	62%	72%	100%	75%

* Calculated effective working time for an 8 hour shift.
** Calculated figures based on the data tabulated above.

Additional notes to the above table.

(1) Hand held equipment — cement grouted rockbolts.

Bolting capacity is primarily limited by the following constraints: rock type (penetration rate), hole length, height above ground, bolt type, operator fitness and skill.

(2) Semi-mechanized equipment — cement grouted rockbolts.

Bolting capacity is primarily limited by the constraints mentioned in (1) above, as well as type and size of rockdrill, availability of antistuck drill automation and speed of boom movement, availability of service platform on rig or on separate unit. Fastest bolting was obtained when two rockdrills were combined with one service platform operated by two men. In the calculation a 1-boom rig is assumed. It is also assumed that each bolt is installed before next hole is drilled.

(3) Fully mechanized equipment — cement grouted rockbolts.

Bolting capacity is primarily limited by the constraints mentioned in (1) above, as well as type and size of rockdrill, availability of antistuck drill steel automation and speed of boom movements. Ceminject grouting system is assumed. The results indicate that one man with a fully mechanized equipment can always set more bolts/shift than he can with semi-mechanized bolting equipment (cement grouted rockbolts). With 2 men on the rig, the bolting capacity will improve by approximately 20%, but at the expense of an extra man. A 2 man operation will simultaneously reduce the capital cost per bolt.

(4) Semi-mechanized equipment — Swellex.

The most rapid bolt installation is obtained by using a standard drill jumbo with a single boom and one operator, plus a separate bolting platform with one man for manual bolt installation. An average installation rate of 71 bolts/shift can be expected.

(5) Fully mechanized equipment — Swellex.

Automation of installation of friction anchored bolts, and particulary Swellex bolts is easier than is the case with grouted rockbolts, since cement or resin handling will no longer be required. The equipment can be made smaller, since only two indexing functions, drilling and bolting, are required. This makes both the design and operation of the machine easier. An extra man will only improve the capacity by approximately 10%, at the expense of one extra man.

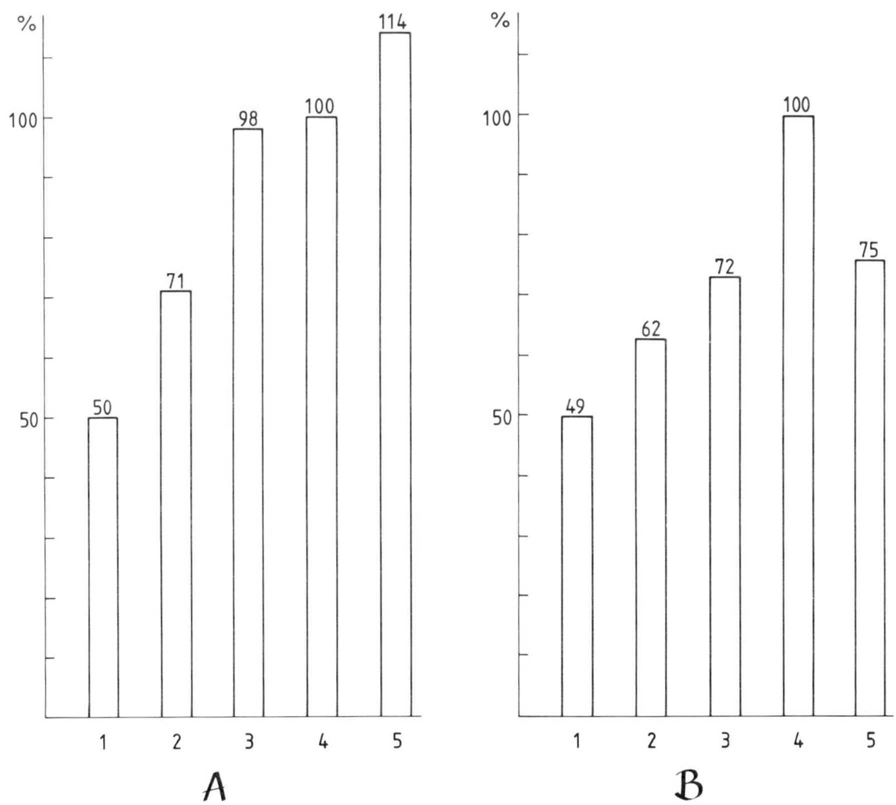

Fig. 44 Bolting capacity relations between different rockbolting equipment.
A. One man operation
B. Two man operation
1. Hand held — cement grouted bolts
2. Semi-mechanized — cement grouted bolts
3. Fully mechanized — cement grouted bolts
4. Semi-mechanized — Swellex
5. Fully mechanized — Swellex

Cost comparison

In a cost analysis the interest rate and the labour cost will affect the result of the analysis significantly. Additional factors of interest are reliability, availability and utilization of the equipment, which may also affect the costs involved significantly.

Pre-set Conditions

New equipment for both alternatives
Semi-mechanized hydraulic equipment
— 1 × 1-boom rig
— 1 × service platform with mortar mixer and pump
Fully mechanized hydraulic equipment
— 1 × Bolting rig with the Ceminject-system

Bolt length: 2.4 m
Bolt diam.: 25 mm
Labour cost US$ 11.06/hour
Tot. labour cost US$ 20.800/year
Working time: 390 shift/year
8 hrs/shift
Depreciation time: 5 years
Interest rate: 16 %

Rock bolting costs in US$ (Feb. 1986 average $ value) for different equipment.

Capital cost	Semi-Mech.	Fully Mech.	Semi-Mech.	Fully Mech.
	Cement grouted rockbolts		Swellex	
Drilling/bolting equipment	240.000	300.000	190.000	220.000
Working platform	60.000		60.000	
Investment cost	300.000	300.000	250.000	220.000
Annuity factor	0.30541	0.30541	0.30541	0.30541
Capital cost/year	91.600	91.600	76.400	67.200
1 man operation				
Bolting capacity				
— per shift pcs*	30	41	42	48
— per year pcs	11.700	15.990	16.380	18.720
Cost/bolt US$	7.83	5.73	4.66	3.59
2 man operation				
Bolting capacity				
— per shift pcs*	44	51	71	53
— per year pcs	17.160	19.890	27.690	20.670
Cost/bolt US$	5.34	4.61	2.75	3.25

(Continued)

	Semi-Mech.	Fully Mech.	Semi-Mech.	Fully Mech.
	Cement grouted rockbolts		Swellex	
Overhauling cost	US $/bolt	US $/bolt	US $/bolt	US $/bolt
Rock drill	0.30	0.30	0.30	0.30
Drill-/bolting rig	1.30	1.67	1.30	1.30
Extra costs incurred by rock fall	0.50	0.25	0.35	0.17
Working platform	0.83	—	0.70	—
Spares stock	0.20	0.23	0.20	0.17
Cost/bolt US $	3.13	2.45	2.85	1.94
Other costs	US $/bolt	US $/bolt	US $/bolt	US $/bolt
Energy costs				
—drill-/bolting rig				
—diesel	0.07	0.07	0.07	0.07
—electricity	0.13	0.10	0.13	0.10
—working platform				
—diesel	0.07	—	0.07	—
Labour cost				
—drill-/bolting				
1 man/2 man operation	3.55/2.42	2.60/4.18	2.53/1.52	2.23/4.03
—working platform				
1 man/2 man operation	—/2.42	—	—/1.52	—
Drillsteel cost	1.33	1.33	1.33	1.33
Material cost				
—bolt	4.00	4.00	7.33	7.33
—face plate	1.00	1.00	1.00	1.00
—cement	0.40	0.40	—	—
Cost/bolt US $	10.55/11.84	9.50/11.08	12.46/12.97	12.06/13.86

(Continued)

	Semi-Mech.	Fully Mech.	Semi-Mech.	Fully Mech.
	Cement grouted rockbolts		Swellex	
Total bolting cost	US $/bolt	US $/bolt	US $/bolt	US $/bolt
1 man operation				
Capital cost	7.83	5.73	4.66	3.59
Overhauling cost	3.13	2.45	2.85	1.94
Other costs	10.55	9.50	12.46	12.06
Total cost	21.51	17.68	19.97	17.59
Cost relations in %	122.3%	100.5%	113.5%	100%
Total bolting cost	US $/bolt	US $/bolt	US $/bolt	US $/bolt
2 man operation				
Capital cost	5.34	4.61	2.75	3.25
Overhauling cost	3.13	2.45	2.85	1.94
Other costs	11.84	11.08	12.97	13.86
Total cost	20.31	18.14	18.57	19.05
Cost relations in %	106.6%	95.2%	97.5%	100%

* Figures taken from table in **Capacity comparison**

Summary cost comparison

One man operation:

Fully mechanized bolting of cement grouted rebar bolts as well as Swellex bolts will cost approximately US$ 3.80 and US$ 2.30, respectively, less per bolt than semi-mechanized rock bolting. The primary reason for this is the higher bolting capacity that can be obtained using fully mechanized equipment in a one man operation. The cost for using Swellex bolts or cement grouted rebar bolts is the same for fully mechanized bolting equipment. Using semi-mechanized equipment, the cost of Swellex is US$ 1.50 lower than that for cement grouted rebars.

Two man operation:

The calculation indicates that the lowest cost per bolt will be obtained using cement grouted bolts in a fully mechanized bolting rig. The most expensive method,

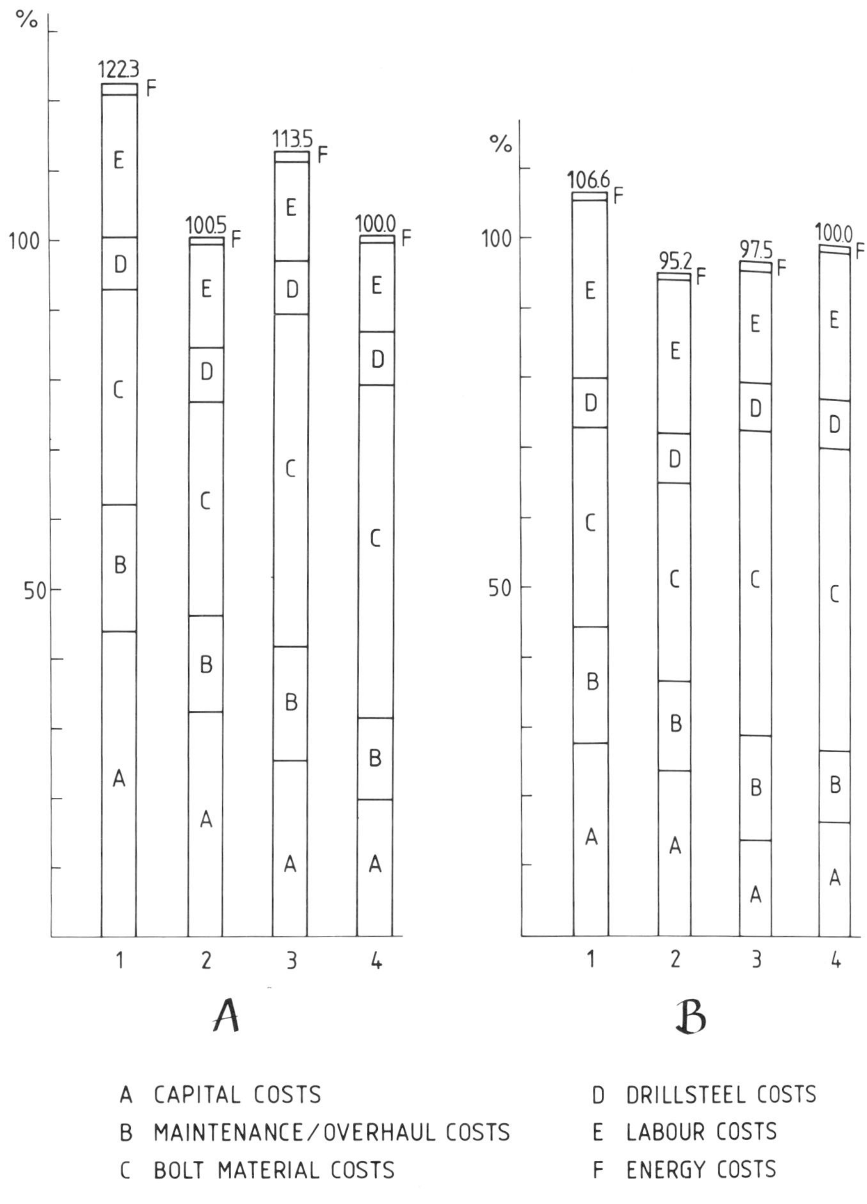

Fig. 45 Bolting cost relations between different rockbolting equipment.
A. One man operation
B. Two man operation
1. Semi-mechanized — cement grouted rockbolts
2. Fully mechanized — cement grouted rockbolts
3. Semi-mechanized — Swellex
4. Fully mechanized — Swellex

however, is the cement grouted rockbolt installed by semi-mechanized bolting equipment. If the same rig is also used for drifting as well as for semi-mechanized bolting, the capital cost is shared and the cost of the bolting will be reduced significantly. The use of Swellex in fully mechanized rock bolting equipment means US$ 0.5—1.0 higher costs per bolt installed. On the other hand, semi-mechanized equipment gives, in a two man operation, a bolting capacity which is superior to any other of the methods compared.

11.4 Development of Rockbolts and Rock Bolting Equipment

Rock reinforcement is a necessary operation in most underground rock excavations. An improved understanding of the mechanism of rock reinforcement will lead to the further development of equipment and methods used. Rockbolting is currently (1986) the most widely used method of rock reinforcement.

Rockbolts and rockbolting methods

When we consider the future of rockbolting, one must not forget that rockbolting is a necessary but disturbing activity in the actual productive work underground. All development will aim principally at minimizing the time and cost of this disturbance. A further development motive is of course the improvement of underground safety and working conditions. Another is the possibility of more advanced excavation geometry.

Cost and time optimization for rockbolting compels the product developers to apply a total system approach to their work. The cost of the rockbolt itself, or its tensile strength is not of so much interest. The total cost and the time needed to reinforce a given rock mass and the safety of the reinforcement system installed is more important. Most of the new types of rockbolts introduced in recent years have been presented, together with a method for rapid and easy installation. Examples of such developments are the Split Set that is installed with the same hand held rock drill that drills the bolthole and the Swellex that is installed by water pumping, without the need of application of any external force to the bolt.

Today's improved understanding of rockbolt action in the rock mass will certainly result in both a need for, and development of more flexible rockbolt systems. These rockbolts will have different properties and behaviour in different reinforcement situations. For example they could be installed to work as a full length supporting anchor in a typical beam forming situation, a sliding bolt in a rock mass where gross deformations are anticipated or a tensioned bolt in a rock mass that has already been subjected to extensive deformation. Such a bolting operation, where every bolt is optimized for its job, will require fewer bolts to be installed to reach the required safety, or provide increased safety for the same bolting effort.

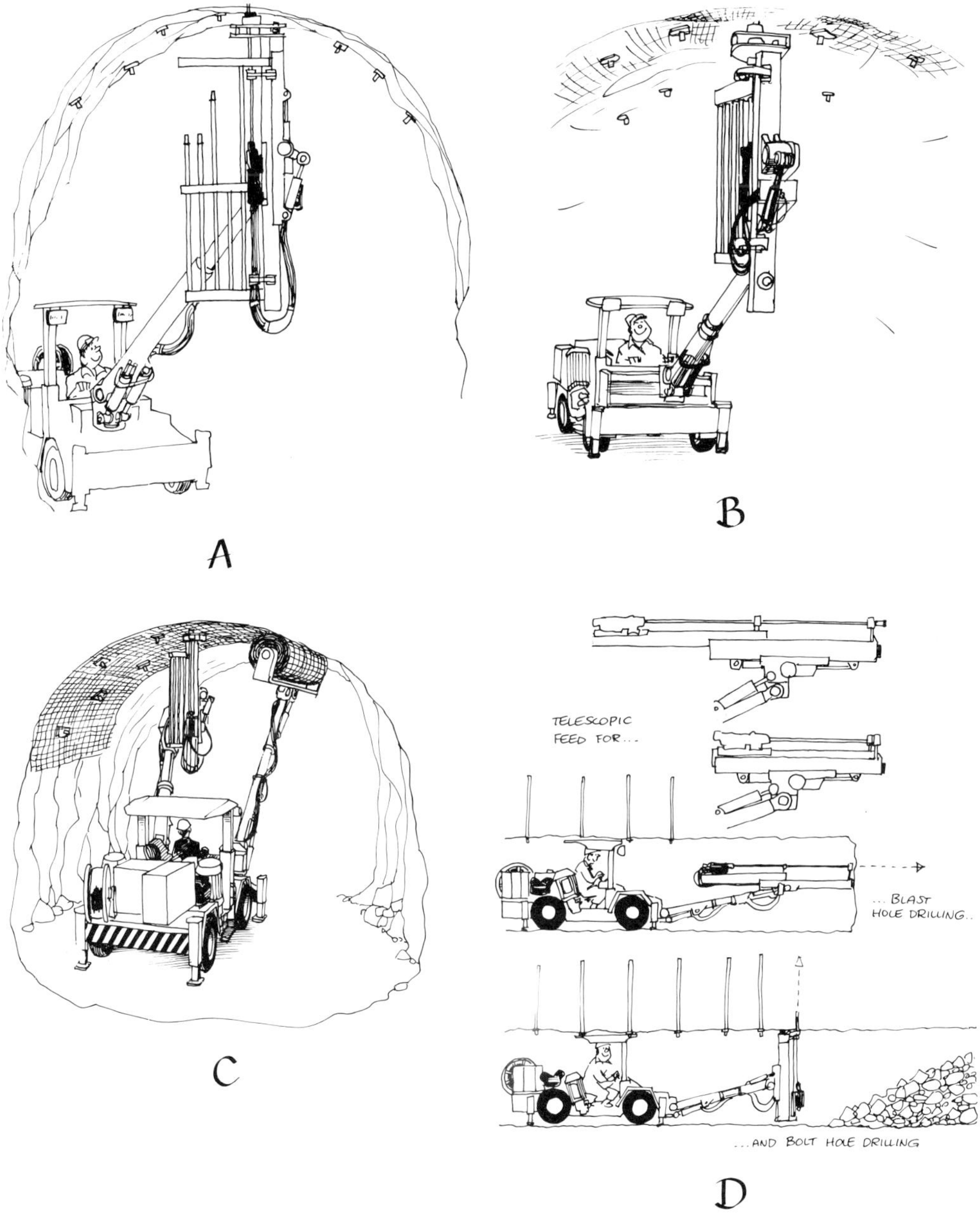

Fig. 46 A—D Schematic illustration of some recent and future trends in mechanization of rock bolting.

A. One bolting rig, different bolt lengths.
B. Fully mechanized bolting rigs for safety.
C. Remote control bolting and wire mesh installation.
D. General purpose rig with telescopic feed for blasthole and bolthole drilling in narrow tunnels.

Increased flexibility could also mean that equipment used for installing bolts could handle different bolt types simultaneously. In the application of the New Austrian Tunnelling Method, different bolt lengths may be required on different positions of the tunnel contour. Today's fully mechanized bolting rigs can handle only one bolt length at a time and require rebuilding to handle another bolt length. Bolting in NATM is often, therefore, a mixture of mechanized and manual bolt installation, since otherwise unnecessarily long bolts must be installed in some positions. More flexible bolting rigs can be expected in the future, see Fig. 46A.

Rockbolting mechanization

From an operator safety point of view, fully mechanized bolting rigs are superior to any other type of rockbolt installation equipment see Fig. 46B. However, only a very small percentage of the bolts installed anually are installed in this manner. Limited flexibility of fully mechanized bolting rigs is one important reason for this. Different bolt lengths must be handled as well as wire mesh, face plates, various attachments for fixation of electric cables, ventilation tubes etc. etc. Complete mechanization will require hand in hand development of the installation machines and the equipment to be installed by them, see Fig. 46C.

A semi-mechanized bolting rig is a cheaper and often also a more versatile solution to the problem of rockbolting mechanization. A drill rig with booms for multi-directional drilling, drills the boltholes. The operator is under a safe protective roof during the drilling phase of the bolting cycle, when the risk of rock fall is highest. Booms can be used for lifting and stretching wire mesh, but the bolt installation is manual. The same rig is used for blast hole drilling. This method is made possible by the development of multi-directional drill booms and telescopic feeds. The telescopic feeds can be operated vertically under limited roof heights, and deep blast hole drilling is also still possible, see Fig. 46D.

Long rockbolts installation

A special problem arizes when the bolt needed is longer than the available space from which bolt installation is to be performed. The most common situation is in low seam room and pillar coal mining. Several development projects, sponsored by the US Bureau of Mines, have lead to technical solutions for full mechanization and the remote control of such a bolting operation, but the projects have never been successful commercially. The safety for the miner has instead been solved by the use of hydraulic temporary roof supports. However, the bolt installation procedure remains the same, with manual bending and restraightening of the bolt. Only a method that is faster and/or cheaper has any chance of replacing the existing technique.

The situation in cable bolting is similar, but here the need for mechanization is urgent because of weight, special cable handling methods and the cement grouting equipment. Cable bolting is one area of bolting where substantial development can be expected. Increased hole drilling precision for the cable will lead to more accurate installation of the reinforcement system. A larger variety of underground excavations may then be developed. Further improvement of the installation equipment will also make cable bolting more competitive with conventional bolting methods in cut and fill mining. Reinforcement of several slices of ore simultaneously gives both a more rational and safer mining operation.

Bolthole drilling

Most rockbolts are installed in 25 to 45 mm diameter holes. In a very few cases, solid steel bolts with diameters of 20 mm or more are used. Most holes are therefore oversized. The reason for this is that small diameter holes are difficult and expensive to drill. Hand held percussive rock drills are best suited to drill 32 to 38 mm diameter holes. We therefore find that may bolts, like expansion shell anchored rockbolts, rebars, Split Set, Swellex etc. are designed to fit this hole size. Rotary drills can effectively drill holes down to about 27 mm in diameter. Bolting machines in coal mines often drill holes of this diameter, which are to be used for 16 to 20 mm rebar rockbolts. Whenever resin grouted bolts are used, methods for drilling even smaller holes (diameter less than 27 mm) are required to minimize the cost of resin. Drilling with high pressure water jets has been tested for such purposes. The water jet, however, is not suitable for all rock types. Sandstone and many hard crystalline rocks can be penetrated by water jet, whereas more "compliant" rocks like limestone can be very difficult to drill using a water jet. Power consumption in water jet drilling is also very high relative to other drilling methods, since, for example, it consumes three to ten times the energy required in hydraulic rotary or percussive drilling. Despite the advantages in very small diameter hole drilling, and its very moderate thrust force and rotation torque requirements, the water jet method will probably be of only limited use in rock bolting. The fastest drilling, mechanized percussive drills, drill holes of roughly 45 mm diameter. In cases where the same drill rig is used for blast hole and bolthole drilling, the hole diameters are often far larger than those required for the bolt. Since the use of this combination of blast hole and bolthole drill rig is increasing, it can be expected that a greater variety of bolts and installation methods for larger size holes will be developed.

Drilling bolts

The ultimate and most efficient bolting method has often been conceived as a method where the bolt drills its own hole and then remains in the rock once the hole is drilled. This development would have to include low cost drill bits for one hole only, the bolt itself as well as methods and equipment for installation. The task might appear to be challenging, but it is of course far from impossible.

List of Figures

Subject Index

Foundations in Tension

— Ground Anchors —

by Prof. Dr. **T. H. Hanna,**
University of Sheffield, England

1982, 600 pp, 345 figs, 55 tables, 920 refs.
price: US$ 58.00

Series on Rock and Soil Mechanics, Vol. 6
ISBN 0-87849-044-2

CONTENTS: 1. Tensile Foundations and Ground Engineering — 2. Soil and Rock Anchors — 3. Anchor Corrosion and Its Protection — 4. Transfer of Load in Anchor and Associated Problems — 5. The Load Testing of Anchors — 6. The Analysis and Design of Anchored Structures — 7. Ground Anchor Use and the Performance of Anchored Structures — 8. Miscellaneous Aspects of Ground Anchors and Their Use.

The subject of ground anchoring is considered by the great majority of engineers to be a recent innovation in the solution of ground problems where uplift or overtuning forces predominate. Consequently there is a built-in conservatism to the use of this attractive ground engineering technique. In part this conservatism is related to the nature of the construction industry, to the lack of authoritative Codes and Design manuals in many countries, to the sensational reporting of occasional failures in the construction press and to the uncertainty of certain phenomena such as the rate of corrosion of steel tendons. The primary purpose of this text is to discuss in detail the design, construction and use of ground anchors.

The book should be of direct interest to the practising Civil, Structural or Mining Engineer who is involved with the design, construction or maintenance of ground structures. Research engineers, teaching stafff and graduate students should also benefit from it.

TRANS TECH PUBLICATIONS
P.O. Box 1254 · D-3392 Clausthal-Zellerfeld
F.R. Germany